T0233925

Gravitational Waves from Coalescing Binaries

Synthesis Lectures on Wave Phenomena in the Physical Sciences

Editor
Sanichiro Yoshida, *Southeastern Louisiana University*

Gravitational Waves from Coalescing Binaries
Stanislav Babak
2020

Gravitational Waves from Coalescing Binaries

Stanislav Babak

ISBN: 978-3-031-01484-0   paperback
ISBN: 978-3-031-02612-6   ebook
ISBN: 978-3-031-00356-1   hardcover

DOI 10.1007/978-3-031-02612-6

A Publication in the Springer series
SYNTHESIS LECTURES ON WAVE PHENOMENA IN THE PHYSICAL SCIENCES

Lecture #1
Series Editor: Sanichiro Yoshida, *Southeastern Louisiana University*
Series ISSN
ISSN pending.

Gravitational Waves from Coalescing Binaries

Stanislav Babak
CNRS

SYNTHESIS LECTURES ON WAVE PHENOMENA IN THE PHYSICAL SCIENCES #1

ABSTRACT

This book is to help post-graduate students to get into gravitational wave astronomy. We assume the knowledge of General Relativity theory, though we will concentrate on the physics and often omit mathematically strict derivations. We provide references to already existing literature where possible, this helps us to see a broad picture, skipping the details. The uniqueness of this book is in that it covers three frequency bands and three major world-wide efforts to detect gravitational waves. The LIGO and Virgo scienti?c collaboration has detected ?rst gravitational waves and the merger of black holes become now almost a routine. We do expect many discoveries yet to come, especially in the joined gravitational and electromagnetic observations. LISA, the space-based gravitational wave observatory, will be launched around 2034 and will be able to detect thousands of GW sources in the milli-Hz band. Pulsar timing array observations have accumulated 20-years' worth of data and we expected detection of GWs in the nano-Hz band within the next decade. We describe the gravitational wave sources and data analysis techniques in each frequency band.

KEYWORDS

general relativity, data analysis, LIGO, LISA, PTA, gravitational waves, black holes, pulsar timing array, analytical relativity, numerical relativity

Contents

CHAPTER 1

Introduction

1.1 FOUNDATION OF GENERAL RELATIVITY

Einstein's General Relativity (GR) theory is currently a widely accepted theory of gravitation. It has passed numerous tests in the solar system and consistency tests based on the observed binary neutron stars and gravitational waves [150]. The gravitation in GR is associated with the Riemannian geometry of spacetime and all matter (including fields) are the sources of the gravitational field. The field equations are nonlinear in the gravitational potentials (associated with the metric of spacetime) and allow exact solutions only under some limited (simplified) assumptions; some of those solutions are given later in this chapter.

We would like to emphasize that this book *is not* a book on GR and we refer the reader to [100, 116] to get more familiar with the theory. The main emphasis of this book is on *Gravitational Waves* (GWs), their sources, and methods of their detection. The GWs are one of the main predictions of GR, they propagate with the speed of light and have two independent polarizations. Having said that, we will introduce some elements of GR in this chapter and the next one, this is the minimum required for the completeness of GWs description.

Throughout this book we adopt the geometrical units $G = c = 1$. Sometimes we restore the speed of light to emphasize its presence, and this will be explicitly stated. We use Greek indices to describe spacetime coordinates and they run over $0, 1, 2, 3$ while the Latin indices dedicated to the spatial coordinates only and they run over $1, 2, 3$. Almost always we work in a certain coordinate frame and we specify the frame we use. The choice of the frame is dictated by what is most convenient in a given problem, and we explicitly mention if the results are independent of the coordinate frame we have chosen. We denote the metric of Minkowski (flat) spacetime as $\eta_{\mu\nu}$.

Next, we give a necessary GR primer which will be needed later for the description of GWs. This will also serve as an introduction of notations used in the book.

The Einstein field equation in geometrical units are given as

$$G^{\mu\nu} = 8\pi T^{\mu\nu},$$ (1.1)

where $G^{\mu\nu}$ is the Einstein's tensor

$$G^{\mu\nu} = R^{\mu\nu} - \frac{1}{2} g^{\mu\nu} R$$ (1.2)

and $T^{\mu\nu}$ is the stress-energy tensor which describes the distribution and motion of the matter. The Ricci tensor $R_{\mu\nu}$ and Ricci scalar R are associated with the Riemann's curvature tensor as

$$R_{\mu\nu} = R^{\alpha}{}_{\mu\alpha\nu}, \quad R = R^{\alpha}{}_{\alpha}, \tag{1.3}$$

where the manipulations with indices are done using the metric $g_{\mu\nu}$ and we always assume summation over the repeated indices. These equations imply that there is no matter which is neutral with respect to the gravitational interaction. This universal coupling is another way to see the principle of equivalence and it allows to formulate the field theory in the geometrical way: the non-flat spacetime geometry is associated (actually equivalent) to the gravitational field sourced by matter. The matter source is, in turn, affected by the gravitational field and its motion can be associated with the trajectories in the non-flat space-time manifold described by the Ricci tensor and the associated metric according to the Einstein's field equations.

Existence of the Bianchi identities

$$R^{\alpha}{}_{\beta\gamma\delta;\epsilon} + R^{\alpha}{}_{\beta\delta\epsilon;\gamma} + R^{\alpha}{}_{\beta\epsilon\gamma;\delta} \equiv 0$$

implies that $G^{\mu\nu}{}_{;\nu} = 0$, where the semicolon ";" defines the covariant derivative with respect to the full metric $g_{\mu\nu}$. This means that not all ten components of the metric can be determined by the Einstein equations, and this can be seen as a freedom in choosing a coordinate frame. Sometimes the choice of the frame is done implicitly by imposing four conditions on the metric tensor. According to the Einstein's equation we also obtain

$$T^{\mu\nu}{}_{;\nu} = 0, \tag{1.4}$$

which is a covariant conservation law. This equation implicitly involves the metric and it implies that we cannot clearly separate matter and gravitational field evolution. Equation (1.4) is also called sometimes the equation of motion for the matter source.

Important role in our further discussion plays the local flatness theorem. This is a mathematical statement which says that at a given point in the curved spacetime we can choose a coordinate frame such that the metric takes Minkowski form and all first derivatives of the metric (and therefore the Christoffel symbols) can be set to zero. If the spacetime is not flat, we cannot set to zero the second derivatives of the metric which is the manifestation of the non-zero Riemann tensor. While the theorem is valid only at a point, we can usually extend it to some finite region of space, of course this extension is only an approximation. How large such an extension can be depends on the tolerance (acceptable accuracy) of smallness in deviation from the flat tangent space and on the radius of curvature at a given point. Usually the characteristic size of that region is given by demand $x/\mathcal{R} \ll 1$, where x is a characteristic size and $\mathcal{R}$ is a radius of curvature (assuming that the curvature does not change much over x). In such a region we can introduce a *local inertial coordinate frame* which is also sometimes referred to as a local Lorentz frame, where effectively the gravitational field is uniform within a desired accuracy.

Later on, we will talk about black holes, so we want to introduce two exact static vacuum solutions of Einstein equations: (i) static and spherical symmetric solution first obtained by K. Schwarzschild and (ii) static axi-symmetric solution found by R. Kerr. The Shwarzschild metric is given by the line element

$$ds^2 = -g_{00}dt^2 + g_{rr}dr^2 + r^2 d\Omega^2, \tag{1.5}$$

where the spherical angular element $d\Omega^2 = d\omega^2 + \sin^2\theta d\phi^2$ and the radial coordinate r is defined so that $2\pi r$ is a circumference. We assumed that the metric is static; that implies that the line element is independent of the direction of time and automatically eliminates g_{0i} components. If we further impose that the metric should be asymptotically flat $g_{\mu\nu} \to \eta_{\mu\nu}$ as $r \to \infty$, then the solution of Einstein equations is given as

$$ds^2 = -\left(1 - \frac{r_g}{r}\right)dt^2 + \left(1 - \frac{r_g}{r}\right)^{-1}dr^2 + r^2 d\Omega^2 \tag{1.6}$$

and the constant of integration r_g is defined from the Newtonian limit: the metric far away from the gravitating body is $g_{00} \approx \eta_{00} + 2\Phi$ where $\Phi = M/r$ is a Newtonian gravitational potential, so $r_g = 2M$. This constant, r_g, is also called a gravitational radius of a body or a *Schwarzschild radius*. Note that the numerical value of Schwarzschild radius depends on a particular choice of the radial coordinate, and it also coincides with the location of an *event horizon*. The event horizon is a global notion and independent of a particular choice of the coordinate frame. It is defined through a congruence of radial null geodesics: none of those geodesics originated from beneath of the event horizon can escape to infinity. Presence of an event horizon defines the *black hole*. The Schwarzschild solution corresponds to a non-spinning black hole (BH) and it is described only by one parameter: its mass. There is plenty of literature dedicated to BHs and properties of the event horizon (e.g., [80, 100, 146]), so we will not describe it here.

The Kerr solution depends on one extra parameter, spin a, which can vary within $[-M^2, M^2]$ and the line element in the Boyer–Lindquist coordinates is given as

$$ds^2 = -\frac{\Delta - a^2\sin^2\theta}{\rho^2}dt^2 - 2a\frac{2Mr\sin^2\theta}{\rho^2}dt\,d\phi +$$
$$\frac{(r^2 + a^2)^2 - a^2\Delta\sin^2\theta}{\rho^2}\sin^2\theta d\phi^2 + \frac{\rho^2}{\Delta}dr^2 + \rho^2\theta^2, \tag{1.7}$$

where

$$\Delta = r^2 - 2Mr + a^2; \quad \rho^2 = r^2 + a^2\cos^2\theta. \tag{1.8}$$

Let us say a few words about this solution: (i) for $a = 0$ the Kerr metric reduces to Schwarzschild; (ii) as $r \to \infty$ the coordinates reduce to the oblate spheroidal; (iii) there is an event horizon (but not at $g_{tt} = 0$ which defines the ergosphere) at $\Delta = 0$: $r_{\text{hor}} = M + \sqrt{M^2 - a^2} \leq 2M$, so this solution is also referred to as a Kerr BH; (iv) metric does not depend on t and ϕ coordinates explicitly, which implies that there are at least two conserved quantities for a test mass motion (according to the Noether theorem): E, L_z which are the energy and projection of the orbital angular momentum on the direction defined by the spin; and (v) $g_{t\phi} \neq 0$, which implies the frame dragging: the test mass dropped radially into a Kerr BH will acquire the angular velocity. We have mentioned that $g_{tt} = 0$ defines the ergosphere (it is also called a "static limit") and its main feature is that no particle could remain at rest inside the ergosphere. Besides two already-mentioned conserved quantities for a test mass moving in the Kerr geometry, there is a third first integral, called *Carter constant*, Q, and it is not associated with the Killing vectors describing the symmetries of the spacetime. We again refer readers to [80, 100, 146] for a more detailed description of black holes.

1.2   GRAVITATIONAL WAVE ASTRONOMY

The first GW signal was detected by laser interferometric gravitational wave observatory (LIGO), these are two interferometers located in the U.S. (one in Hanford, WA, another in Livingston, LA) on September 14, 2015 [4]. The event was detected by both LIGO detectors within a light travel time, and the wave matched the predictions of general relativity for a gravitational wave produced by a system of two merging stellar-mass BHs. The first detection, GW150914, was a triple lucky event: (i) the detectors were operational in the science mode; (ii) it is the loudest GW signal from merging BHs detected in the first observing run; and (iii) it falls in the part of the parameter space where we are very confident about the theoretical model of GW signals used in the search. A few months later, the detection of a second source, GW151226, was also announced [3]. Soon the field will reach a milestone with the construction of an international network of GW interferometers that have achieved (or are close to) their designed sensitivity. The LIGO-Virgo network will be extended in the near future with new advanced detectors such as a cryogenic underground detector in Japan and another LIGO-type detector in India. The extended network can increase, by factors of 2 to 4, the detected event rate by performing coherent data analysis. The coincidence of the true GW signals across the detectors allows reducing the spurious instrumental background caused by the non-stationarity of the noise. Currently, three detectors are operating simultaneously (two LIGO and VIRGO). The network of detectors not only increases the signal-to-noise ratio of a signal, but also allows better estimation of parameters, especially measurement of both polarizations and localization of the source in the sky, enabling the electromagnetic follow-up observations.

In this book, we will primarily concentrate on the binary systems as GW sources and we only briefly mention other sources. There are already quite a few books on gravitational waves, and some of them give better description of detection or modeling GW signals. The distinct feature of this book is that we will get through three major efforts to detect GWs in different frequency bands: few Hertz to kilo-Hertz with LIGO-like detectors, 0.1–100 mHz with the future space-based detector LISA, and the use of millisecond pulsars (Pulsar Timing Array) to detect GWs in the nano-Hertz band. GW at the ultra-low frequencies could be indirectly detected by observing the *B*-polarization mode of the cosmic microwave background (see, for example, [11]) and it will not be discussed here.

Almost at the same time as the first GW detection, the small space mission LISA Pathfinder delivered fantastic results [17]. LISA Pathfinder, a proof of concept for the future LISA (Laser Interferometer Space Antenna) GW observatory, was launched in December 2015. The results were beyond all expectation, they were better than minimum requirement by a factor 10–1000 (depending on the frequency), achieving a sub-Femto-*g* in free fall. This small mission has tested the most critical technologies and demonstrated the readiness required to build a large scale GW observatory in space. LISA has been scheduled for launch by the European Space Agency around 2034.

Three major worldwide Pulsar Timing Array (PTA) collaborations (NANOGrav (North America), PPTA (Australia), and European PTA) are working together to detect gravitational waves in the nano-Hertz band. There is no detection yet, but the obtained upper limits on the strain of GWs allows us to rule out some most optimistic astrophysical models.

In this book we consider the basis for modeling GW signals from binary systems. Then we will discuss the data analysis methods used for detecting GWs. Note that those methods have much wider applications, and we will specify how they apply to GW astronomy. In the last chapter we give details of the major GW sources and the data analysis methods specific to each frequency band and technology used in the search for GWs.

CHAPTER 2

Gravitational Waves from Coalescing Binaries

In this chapter, we will cover the generation of gravitational waves and their main properties. We restrict our attention to the GWs from the coalescing binary systems. We start with the leading order description and then discuss the post-Newtonian approximation. Next, we describe the extreme mass ratio inspirals (EMRIs) and the basis for the effective-one-body approach. Finally, we touch the numerical relativity and its role in the GW signal modeling.

The description of linearized GWs is closely follows set of lectures given by Prof. K. Thorne in Caltech and in Ecole d' Été de Physique Theorique (Les Houches, France), 1966, 1972, and 1982.

2.1 WEAK GRAVITATIONAL WAVES, LINEARIZED THEORY

In this section we introduce a weak gravitational wave in vacuum. We start with a frame invariant description manipulating with the Riemann tensor, followed by GW description and solution for the metric. We will show that the GWs in GR are transverse and traceless and have two independent polarizations.

Consider weak gravitational waves, that is we consider only a leading order in the GW amplitude contribution. We also assume that $\lambda^{GW} \ll \{\mathcal{R}, \mathcal{L}\}$, where λ^{GW} is a typical GW wavelength, $\mathcal{R}$ is a curvature of background geometry $\mathcal{R} \sim 1/\sqrt{R_{\alpha\beta\mu\nu}}$, and $\mathcal{L}$ is a characteristic scale on which $\mathcal{R}$ changes. The assumption $\lambda^{GW} \ll \mathcal{R}$ implies that the size of the source (which characteristic size is of the same order as GW wavelength) is much smaller than the curvature of the external Universe. This assumption is broken if the GWs belong to the stochastic GW background of cosmological origin. There the GW wavelength could be comparable or even larger than the Hubble radius; this leads to a very efficient interaction between the expanding Universe and the long-wavelength GWs [74]. We also consider here GWs far away from the source (far zone: distance to the source much larger than λ^{GW}), so that we do not deal with the curvature created by a source itself. In all other cases the above assumption holds.

We want to separate the background geometry (created by gravitational field of galaxies, solar system, cosmology, etc.) and GWs. To do so, we define the "background" Riemann tensor as the total Riemann curvature averaged over the scale larger than several GW wavelengths:

$$R^B_{\alpha\beta\gamma\delta} = \langle R_{\alpha\beta\gamma\delta}\rangle; \tag{2.1}$$

the oscillations due to GW average out and we should have a smooth component of the background. Now we can introduce the contribution due to GWs:

$$R^{GW}_{\alpha\beta\gamma\delta} = R_{\alpha\beta\gamma\delta} - R^B_{\alpha\beta\gamma\delta} \sim \frac{h}{\lambda^2}, \tag{2.2}$$

where h is a characteristic strain of GW. Note that in all cases which we consider in this book

$$R^B_{\alpha\beta\gamma\delta} \sim \frac{1}{\mathcal{R}^2} \ll R^{GW}_{\alpha\beta\gamma\delta} \sim \frac{h}{\lambda^2}, \tag{2.3}$$

one can easily check it by using, for example, the Hubble radius as $\mathcal{R}$ and the strain and wavelength of the detected by LIGO GW signals. Prof. K. Thorne compares the GWs on the smooth background with the little bumps on the skin of an orange: GWs are much smaller than the orange itself but the curvature of those bumps is larger than the curvature of an orange.

Similarly, we can also split the metric into background and GW:

$$g_{\mu\nu} = g^B_{\mu\nu} + h^{GW}_{\mu\nu}; \quad g^B_{\mu\nu} = \langle g_{\mu\nu}\rangle. \tag{2.4}$$

We denote covariant derivatives with respect to the background as "$|\mu$" and keep the semi-colon for the total covariant derivative. Take the second derivative of the GW Riemann tensor w.r.t. background metric:

$$R^{GW}_{\alpha\beta\gamma\delta|\mu}{}^{|\mu} \,\&\, R^{GW}_{...} R^B_{...}. \tag{2.5}$$

The sign "&" means that the r.h.s. is constructed out of the terms of this type without giving exact expression. The right-hand side describes coupling of the GWs and the background, however

$$R^{GW}_{...} R^B_{...} \sim \frac{R^{GW}}{\lambda^2}\frac{\lambda^2}{\mathcal{R}^2}$$

which implies that the r.h.s. can be neglected as it is of higher order (factor $\sim \lambda^2/\mathcal{R}^2$) than the l.h.s. and we get (in the leading order)

$$R^{GW}_{\alpha\beta\gamma\delta|\mu}{}^{|\mu} = 0. \tag{2.6}$$

We can simplify our life even further if we introduce local (inertial) Lorentz frame. Indeed, far away from the gravitational field sources the background spacetime can be described as Minkowski with a very high precision. In other words, there is such a scale $\lambda^{GW} \ll l \ll \mathcal{R}$, where we can approximate the background metric as Minkowski. Note that the "local" in here could describe quite a large range where we can introduce a local inertial frame (LIF) with respect to the background spacetime:

$$R^{GW}_{\alpha\beta\gamma\delta,\mu}{}^{,\mu} = \Box R^{GW}_{\alpha\beta\gamma\delta} = 0, \tag{2.7}$$

where the comma defines usual partial derivatives and the box is a flat D'Alembert wave operator which is given in Cartesian coordinates as

$$\Box = -\frac{\partial^2}{\partial t^2} + \frac{\partial^2}{\partial x^2} + \frac{\partial^2}{\partial y^2} + \frac{\partial^2}{\partial z^2}.$$

Equations (2.6, 2.7) are the wave-like equations which describe propagation of Riemann tensor in the arbitrary/flat background.

For simplicity, we start with a plane GW solution propagating in z-direction, which implies $R^{GW}_{\alpha\beta\gamma\delta} = R^{GW}_{\alpha\beta\gamma\delta}(t - z)$. We consider the spacetime outside the source, so that we work in vacuum, this implies that the spacetime is Ricci-flat: $R_{\alpha\beta} = R^{\mu}{}_{\alpha\mu\beta} = 0$ (those are Einstein's equations). The only non-trivial components of the Riemann tensor are

$$R^{GW}_{0x0x} = -R^{GW}_{0y0y} = -\frac{1}{2}\ddot{h}_+(t - z) \tag{2.8}$$

$$R^{GW}_{x0y0} = R^{GW}_{y0x0} = -\frac{1}{2}\ddot{h}_\times(t - z), \tag{2.9}$$

where the right-hand side is so far just a notation which will become clear later, it is proportional to the second time derivative (the over dot) of the GW strain. From this we can make two conclusions: (1) all z components of the Riemann tensor are zero, which implies that the GW is a *transverse* wave, it has only components orthogonal to the propagation direction; and (2) the trace is zero, so the GW is *traceless*.

So far, we were working directly with Riemann tensor, which is indicator of the curved spacetime and cannot be eliminated by any choice of coordinate frame. Now we re-work the GW in terms of the metric which has close connection to the Newtonian gravitational potentials. We again use a local inertial (Lorentz) frame where the background metric can be approximated as Minkowski:

$$g_{\alpha\beta} = \eta_{\alpha\beta} + h_{\alpha\beta} + O(h^2); \tag{2.10}$$

here $\eta_{\alpha\beta}$ is Minkowski metric. We claim that there is a specific coordinate frame (called "TT-gauge," where TT stands for transverse and traceless) where only non-zero components of metric are

$$h_{ij} = h^{TT}_{ij} = \begin{pmatrix} h_+ & h_\times & 0 \\ h_\times & -h_+ & 0 \\ 0 & 0 & 0 \end{pmatrix}. \tag{2.11}$$

What is a "gauge"? In terms of GR choosing a gauge is completely equivalent to the choice of a specific coordinate frame. This choice is incorporated in the Einstein's equations through the Bianchi identities and $G^{\mu\nu}{}_{;\nu} = 0$ which implies that only six equations are independent. But it is a bit trickier in terms of the metric decomposition to background and GWs given by Eq. (2.10); there we fix the coordinate frame so that the background has Minkowski diagonal form and we have freedom to modify $h_{\mu\nu}$ without changing the field equations. Indeed, one can verify that the linearized Riemann tensor

$$R^{GW}_{\alpha\beta\mu\nu} = \frac{1}{2}\left(h_{\alpha\nu|\beta\mu} + h_{\beta\mu|\alpha\nu} - h_{\alpha\mu|\beta\nu} - h_{\beta\nu|\alpha\mu}\right) \tag{2.12}$$

does not change under the gauge transformation

$$h_{\alpha\beta} \rightarrow h_{\alpha\beta} - \xi_{\alpha,\beta} - \xi_{\beta,\alpha}, \tag{2.13}$$

where ξ^{α} is an arbitrary vector field which keeps $|h_{\alpha\beta}| \ll 1$ and we have used LIF associated with the background. This is analogous to the gauge transformation in the electormagnetism where the electric and magnetic fields, written in terms of potentials

$$\vec{E} = -\nabla V - \frac{\partial \vec{A}}{\partial t}, \quad \vec{B} = \nabla \times \vec{A},$$

are invariant under the gauge transformation $A^{\mu} \rightarrow A^{\mu} + \nabla^{\mu} f$, where $A^{\mu} = \{V, \vec{A}\}$ and f is an arbitrary scalar field. In order to show that the metric indeed satisfies the wave-like equation, we first need to introduce a trace-reverse metric

$$\bar{h}^{\mu\nu} = h^{\mu\nu} - \frac{1}{2}\eta^{\mu\nu}h, \quad h = \eta_{\mu\nu}h^{\mu\nu} \tag{2.14}$$

and choose the *harmonic gauge* condition

$$\bar{h}^{\mu\nu}{}_{,\nu} = 0, \tag{2.15}$$

then the linearized Einstein equations in vacuum reduce to $\Box\bar{h}^{\mu\nu} = 0$. Note that the use of the harmonic gauge is not essential to get the wave-like solution (as we have shown that by operating with the Riemann tensor directly), however it simplifies the equations and their interpretation. Harmonic condition given by Eq. (2.15) defines the class of the harmonic coordinates among which we can still can choose a particular one. Indeed, the GW strain can be further modified as

$$\bar{h}^{\mu\nu}_{new} = \bar{h}^{\mu\nu} - \xi^{\mu,\nu} - \xi^{\nu,\mu} + \eta^{\mu\nu}\xi^{\alpha}{}_{,\alpha} \tag{2.16}$$

so that the field equations and the gauge condition remains the same if ξ^{μ} satisfies $\Box\xi^{\mu} = 0$. This choice corresponds to four additional degrees of freedom which allows us to choose

$$\bar{h}^{0\alpha} = 0, \quad \bar{h} = 0; \tag{2.17}$$

note that the first condition contains only three independent equations (due to presence of the generic harmonic gauge condition). A more general condition would be $u^{\mu}\bar{h}_{\mu\nu} = 0$, where u^{μ} is a time-like constant vector field. Consider a plane wave solution of the wave equation

$$\bar{h}^{\mu\nu} = A^{\mu\nu}e^{ik_{\alpha}x^{\alpha}}, \quad k^{\alpha} = \{\omega, \vec{k}\}. \tag{2.18}$$

The field equations imply

$$k_\alpha k^\alpha = 0, \tag{2.19}$$

that is the GW is null and propagates with the speed of light, ω is the GW frequency, and $\vec{k}$ is the wave vector defining direction of wave propagation. Other two conditions translate to

$$A_{\mu\nu}k^\mu = 0, \quad A^\mu{}_\mu = 0, \tag{2.20}$$

which again emphasize transverse and traceless nature of GW. Note that there is also a non-radiative solution of vacuum equations, which is for slowly moving source (or static) leads to the Newtonian potential.

To get a "TT" (transverse-traceless) part of the metric (radiative) we can use/apply the TT-operator which gives us the gravitational wave in TT-gauge:

$$h_{jk}^{TT} = \Pr_{jklm} h^{lm} \tag{2.21}$$

$$\Pr_{jklm} = P_{jl}P_{mk} - \frac{1}{2}P_{jk}P_{lm}, \quad P_{jk} = \delta_{jk} - n_j n_k, \tag{2.22}$$

where $n^j = k^j/|\vec{k}|$ is a unit vector in the direction of propagation.

Substituting the found GW metric into the Riemann tensor and working in linear order in h we get

$$R_{0j0k}^{GW} = -\frac{1}{2}\ddot{h}_{jk}^{TT}. \tag{2.23}$$

Last (but not least) in this section, we introduce a polarization basis. Indeed, as we have shown before, in TT-gauge, the GW strain has non-zero components in the plane orthogonal to the direction of propagation. We can introduce the ortho-normal *polarization basis* on that plane: $\hat{e}_1$, $\hat{e}_2$ which satisfy $\hat{e}_1.\hat{e}_2 = 0$, $\hat{e}_1.\vec{k} = \hat{e}_2.\vec{k} = 0$ and define the polarization tensors (using the outer product $\otimes$):

$$\varepsilon^+ = \hat{e}_1 \otimes \hat{e}_1 - \hat{e}_2 \otimes \hat{e}_2 \tag{2.24}$$

$$\varepsilon^\times = \hat{e}_1 \otimes \hat{e}_2 + \hat{e}_2 \otimes \hat{e}_1. \tag{2.25}$$

The GW strain can be then decomposed as

$$h_{ij} = h_+\varepsilon_{ij}^+ + h_\times\varepsilon_{ij}^\times, \tag{2.26}$$

$$h_+ = \frac{1}{2}h_{ij}\varepsilon_+^{ij}, \quad h_\times = \frac{1}{2}h_{ij}\varepsilon_\times^{ij}. \tag{2.27}$$

There is an arbitrariness in the choice of the polarization basis $\hat{e}_1, \hat{e}_2$, but they all related by a rotation angle ψ which is called *polarization angle*. One more important conclusion we can draw from here is that GW has two polarizations h_+, $h_\times$ and they go into each other by rotation $\pi/4$. This follows from the rotation matrix applied to each basis vector that gives us the double angle in the polarization tensor.

2.1.1   PLANE GRAVITATIONAL WAVES AND GEODESIC DEVIATION

In this section we will consider the geodesic deviation in the field of GWs, which will set the foundation for detecting GWs.

Let us remember that the geodesic equation reads as $u^\alpha u^\beta{}_{;\alpha} = 0$, where u^α is a 4-velocity. The geodesic deviation equation tells us about behavior of two nearby geodesics—in the absence of any gravitational field (or equivalently, in absence of the curvature), geodesics are two parallel lines, and any change in the distance between them is governed by a Riemann curvature tensor according to

$$\frac{d^2\xi^\alpha}{d\tau^2} = R^\alpha{}_{\mu\nu\beta} u^\mu u^\nu \xi^\beta, \tag{2.28}$$

where ξ^α is a distance between two geodesics, τ is an affine parameter running along each geodesic (one can see it as a proper time of an observer moving along geodesic), and u^α is a 4-vector tangent to a geodesic (again, can be seen as a 4-velocity of an observer). We apply this equation to the case of a plane GW traveling in z-direction. Consider two particles/observers: the first one (call it "A") is at rest at the origin of a coordinate frame while the second one (call it "B") is nearby, initially separated by a distance $|\xi_0|$.

We assume that the observer A synchronized its clock with the observer B. Both observers are traveling along their geodesics. We can again introduce a local inertial frame associated with the observer A, and, by choosing ξ_0 sufficiently small, we can ensure that our frame covers both observers. The observer A remains at rest in the center of the coordinate frame. We made a little sketch describing the setup in the left panel of Figure 2.1. We use an expression for the Riemann tensor for the plane GW wave obtained in the previous section and we neglect all other possible interaction between two observers. Then the geodesic deviation equation becomes

$$\frac{d^2\xi^i}{dt^2} = R^i{}_{00j}\xi^j = -R^i{}_{0j0}\xi^j$$

$$\frac{d^2\xi^i}{dt^2} = \frac{1}{2}\ddot{h}^{TT}_{ik}\xi^k. \tag{2.29}$$

Note that the indices lowered and raised here with the help of Minkowski metric; as usually, we assume weak GW ($h \ll 1$ so we work in linear approximation). Introduce the deviation in the distance between two observers $\delta\xi^i = \xi^i - \xi_0^i$, then we have the following equation for the deviation:

$$\delta\xi_i = \frac{1}{2}h^{TT}_{ik}\xi_0^k. \tag{2.30}$$

Now instead of two observers, consider a ring of particles (observers) all at the equal distance $|\xi_0|$ away from the observer "A" at the initial time. The equation of motion for each particle is described by Eq. (2.30). The forces acting on the ring of particles are similar to the time-varying tidal force; the circle is deformed into an ellipse. The acceleration is divergence-free so we can introduce the lines of force; they are drawn in the right panel of Figure 2.1.

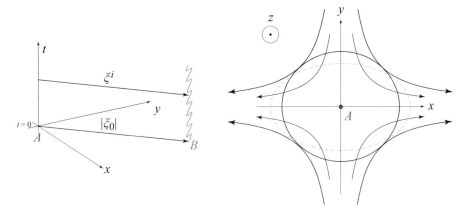

Figure 2.1: Left plot: sketch showing spacetime diagram of two observers, A and B. Right plot: lines of force (tidal) acting on the ring of particles.

If we have a monochromatic and linear polarized GW then in the first quarter of period the circle is squeezed along y direction and stretched along x, then during the second quarter the ellipse regains its circular shape, then it is squeezed in x direction and stretched along y, in the last quarter of period the ring returns into its original shape. We have described here +- polarization; the ×-polarization can be obtained by rotating the picture by 45°. Equation (2.30) gives us the basis for detecting GWs; we attempt to measure the change in the distance between two nearby objects caused by passing GW. The expected GW strain is $h \sim 10^{-23} - 10^{-20}$ so we need a very sensitive device which is also well insulated from any environmental influence. In addition, as follows from Eq. (2.30), the deviation is proportional to the characteristic size of the measuring device ξ_0; this is the reason for using kilometer-long ground-based detectors and making the space-based GW observatory of the size few $\times 10^9$ meters.

2.2   GENERATION OF GRAVITATIONAL WAVES

In this section we consider the generation of GWs from *isolated* sources with *weak* internal gravity. We will work only in the leading order and the results that given here are also available in several textbooks; we recommend the following [91, 100, 116]. We are dealing here with an *isolated* source which means that there is a sphere of a finite radius which completely encompasses the sources. We also demand that the size of the sphere is significantly smaller than the distance to the observer $|\vec{R}|$ (basically we demand that the observer is in the far zone of the source). We have schematically drawn the source in Figure 2.2. The characteristic size of the emitting system is L, the primed coordinates (x', t') denote the points inside the isolated system, and (x, t) are the filed point of the observer(s). During the derivation we will explicitly spell out the assumptions and approximations.

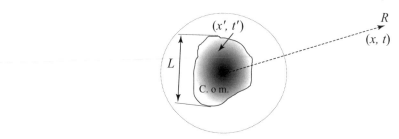

Figure 2.2: Drawing explaining the isolated source of characteristic size L with the center of mass located at the position "c. o m." The observer is at the distance R away from the source.

We assume that the observer is far away, so that we can use local inertial frame (with respect to the background) in its vicinity and the metric given by $g_{\alpha\beta} = \eta_{\alpha\beta} + h_{\alpha\beta}$. As before, we use the trace-reverse strain $\bar{h}^{\alpha\beta}$, and the indices are lowered and raised with the flat metric $\eta_{\alpha\beta}$.

We use harmonic gauge $\bar{h}^{\mu\nu}{}_{,\nu} = 0$, so the Einstein equations become

$$\Box\bar{h}^{\mu\nu} = -16\pi(T^{\mu\nu} + t^{\mu\nu}). \tag{2.31}$$

Here, we have restored the matter term on the r.h.s. (stress-energy tensor $T^{\mu\nu}$) and we have also combined all nonlinear (in h) terms, moved them to the right and denoted by $t^{\mu\nu}$. We purposely used a different color: first of all, it is not a tensor, sometimes it is called "gravitational stress-energy pseudo-tensor;" second, we will neglect this term for now (in the leading order) and restore it later. This term is important as it tells us that the gravitational field becomes its own source, but this effect is of a higher order. This also can be seen as we neglect its contribution as compared to the matter $T^{\mu\nu}$. The formal solution of this equation can be written as

$$\bar{h}^{\mu\nu} = 4\int \frac{T^{\mu\nu}(x', t' = t - |\vec{x}' - \vec{x}|)}{|\vec{x}' - \vec{x}|}d^3\vec{x}'; \tag{2.32}$$

note that the stress-energy tensor is evaluated at the retarded time $(t - |\vec{x}' - \vec{x}|)$.

We assume slow motion of the source $v \ll 1$, where v is a characteristic velocity of the system (remember $c = 1$, so we compare the velocity to the speed of light)

$$v \sim \frac{L}{P} \sim \frac{L}{\lambda^{GW}} \ll 1, \tag{2.33}$$

where P is a characteristic time scale of the evolving system (say, period), and the last inequality implies that the source is inside the near zone. As mentioned at the beginning of this section, we also assume the weak internal gravity of the system. Consider an example where $T^{\mu\nu}$ describes a perfect fluid

$$T^{\mu\nu} = (\rho + p)u^\mu u^\nu + pg^{\mu\nu}, \tag{2.34}$$

where ρ is density and p is a pressure. Due to the weak internal field assumption $p/\rho \sim v^2$, indeed the pressure is, roughly speaking, a force per unit area which is proportional to the momentum flux that is of order ρv^2. We have the following hierarchy among the components of the stress-energy tensor:

$$T^{00} \sim \rho \gg T^{0i} \sim \rho v \gg T^{ij} \sim \rho v^2. \tag{2.35}$$

We have assumed that the observer is in the far zone:

$$|\vec{x} - \vec{x}'| \approx R \gg \lambda^{GW}. \tag{2.36}$$

GWs do not change much while propagating through the system, so the retarded time can be approximated as $t' = t - R$.

We are not interested in the gravitational potential of the system (its almost Newtonian) but want to identify the radiative degrees of freedom, which are described by "TT" part of the gravitational field. Taking all these into account we get

$$h_{jk}^{TT} = \left[\frac{4}{R} \int T_{jk}(x', t' = t - R) d^3 x' \right]^{TT}. \tag{2.37}$$

We can compute the TT-part using the projection operator introduced above (see Eq. (2.22).

Next, we employ the conservation laws: $T^{\mu\nu}{}_{,\nu} = 0$, in particular, it can be shown that

$$T^{jk} = \frac{1}{2} T^{00}{}_{,00} x^j x^k + \left[T^{mk} x^j + T^{mj} x^k \right]_{,m} - \frac{1}{2} \left(T^{mn} x^j x^k \right)_{,mn}. \tag{2.38}$$

When integrating over the volume occupied by a system, the full divergences can be reduced to the surface integrals (over the surface completely surrounding the system, as, for example, shown in Figure 2.2). Those surface integrals are all zero as no matter is present on the surface. The wavelike solution can be written as

$$h_{jk}^{TT} = \left[\frac{2}{R} \frac{d^2}{dt^2} \mathcal{M}_{jk}(t - R) \right]^{TT}, \tag{2.39}$$

where $\mathcal{M}^{jk}$

$$\mathcal{M}^{jk} = \int T^{00} \left(x^j x^k - \frac{1}{3} \delta^{jk} r \right) d^3 x = \left(I^{jk} \right)^{traceless} \tag{2.40}$$

is the mass quadrupole moment and I^{jk} is the second moment of the mass distribution

$$I^{jk} = \int T^{00} x^j x^k \, d^3 x. \tag{2.41}$$

The expression (2.39) gives a leading order (quadrupole) expression. To generate GWs the system should have non-zero second derivative of the quadrupole moment. Besides leading order (quadrupole moment), there are other moments which contribute to the GWs. There are two types of moments: mass moments $I_l \sim M l^l$ and current moments $S_l \sim M v L^l \sim v I_l$. In general, the expression for the GW strain could be written as

$$h_{+,\times} \sim \frac{1}{R} \left[\frac{\partial^2 I_2}{\partial t^2} \& \frac{\partial^3 I_3}{\partial t^3} \& \ldots \& \frac{\partial^2 S_2}{\partial t^2} \& \frac{\partial^3 S_3}{\partial t^3} \ldots \right]; \tag{2.42}$$

as before, we do not give here the exact expression just outline what kind of terms are present, the exact expression one can find in the very nice review by Blanchet [32]. As you can see, the lower multipoles do not contribute to the radiation as they are related to various conservation laws: $I_0 \sim M$ – conservation of mass, $\dot{I}_1$ – linear momentum, and $\dot{S}_1$ – angular momentum. There is a hierarchy in how much each term contributes to the GW strain:

$$\frac{1}{R} \frac{\partial^l I_l}{\partial t^l} \sim \frac{M}{R} v^l, \quad \frac{1}{R} \frac{\partial^l S_l}{\partial t^l} \sim \frac{M}{R} v^{l+1}; \tag{2.43}$$

the dominant contribution comes from the lower multipoles (since $v \ll 1$). If the second derivative of the mass quadrupole moment is zero, the next sub-leading term contains current quadrupole, explicitly

$$\hat{h}_{jk}^{TT}|_{\text{cur.quad}} = -\frac{8}{3R} \left[\epsilon_{jil} n_i \ddot{S}_{lk}(t - R) \right]^{STF}, \tag{2.44}$$

where ϵ_{ijl} is the Levi-Civita anti-symmetric symbol, $n^i = x^i/R$ is a unit vector, and "STF" means symmetric and trace-free part. The current quadrupole moment is given explicitly as

$$S_{jk} = \int \left[\vec{x} \times \rho\vec{v} \right]_j x_k \, d^3x$$

which is an analog of magnetic momentum. Let us have another look at the quadrupole expression Eq. (2.39): $\frac{1}{R} \frac{\partial^2 I_l}{\partial t^2} \sim \frac{M}{R} v^2$, the GW strain depends on the distance (actually more precisely on the luminosity distance) to the source, on the mass of the system and characteristic velocity in the system. The most suitable candidates are various binary systems with the masses similar or larger than solar mass, and to achieve a high velocity we need to have compact companions in the binary. The strongest sources, therefore, are BH binaries. They are the most compact and could be very massive. Solar mass BHs are the final product of the stellar evolution, however to get BH we need rather massive stars (with a mass larger than ~ 15 solar mass) and those are significantly more rare as compared to less massive stars. Other compact binary systems could consist of neutron stars and/or white dwarfs. We describe the GW sources in greater details in the last chapter of this book; the main message which we want to deliver here is that the most promising sources of GWs are various compact (in size and in separation) binary systems of astrophysical origin. The main focus of this book is gravitational radiation from the *coalescing* binaries and its detection. We need to explain the word "coalescing"—the gravitational radiation carries the energy and angular momentum from the system—and, as a result, the binary system shrinks and becomes more circular (if it was eccentric at some point of its evolution).

Consider Einstein equations up to second order in GW strain expansion and far away from the emitting system:

$$G_{\alpha\beta} = G_{\alpha\beta}^B + G_{\alpha\beta}^{(1)} + G_{\alpha\beta}^{(2)} = 0. \tag{2.45}$$

We can separate each Einstein tensor into a smooth ($< \ldots >$) part and in the part which fluctuates (as before, $G_{\alpha\beta}^{(\ldots)} - < G_{\alpha\beta}^{(\ldots)} >$). We can consider the GW far away from the source (vacuum) and equate the smooth parts (up to second order):

$$G_{\alpha\beta}^B = \langle G_{\alpha\beta} \rangle = - < G_{\alpha\beta}^{(2)} > = 8\pi T_{\alpha\beta}^{GW}. \tag{2.46}$$

The right-hand side of the last equation could be associated with the stress-energy tensor for GWs, however it is a bad definition as it tells that GWs are the source of the background, which contradicts our original assumption. Nevertheless, it gives a valid expression for the energy and angular momentum carried by GWs [83]:

$$T_{\alpha\beta}^{GW} = \frac{1}{16} \langle h_{+,\alpha} h_{+,\beta} + h_{\times,\alpha} h_{\times,\beta} \rangle, \tag{2.47}$$

where, as before, the triangular brackets denote averaging over several gravitational wavelengths. The loss of energy and of angular momentum by GWs is given explicitly by

$$\frac{dE}{dt} = -\frac{1}{5} \langle \dddot{\mathcal{M}}_{ij} \dddot{\mathcal{M}}_{ij} \rangle, \tag{2.48}$$

$$\frac{dS_j}{dt} = -\frac{2}{5} \epsilon_{jkl} \langle \ddot{\mathcal{M}}_{ki} \dddot{\mathcal{M}}_{li} \rangle. \tag{2.49}$$

The orbital evolution of a binary system can be obtain by assuming adiabatic evolution: the extraction of energy from a system (inspiral time scale) is much longer than the orbital period, in other words, the rate of change of the orbital energy and of the orbital angular momentum can be equated to the rate their extraction by GWs (given by Eqs. (2.48, 2.49). This technique is used for evolving binary system (*inspiral*) until the adiabatic condition breaks down, which is usually referred to as a *merger* where two objects start to form a common strongly distorted single object.

Finally, we recall the nonlinear term $t^{\mu\nu}$ in the Einstein equation; we have neglected it so far as it is of a higher order and much smaller than the matter source. This term implies that the gravitational field starts to be its own source. The field equations can be solved iteratively and will discuss this in the following sections of this chapter (see Section 2.4).

2.3   CIRCULAR BINARY SYSTEM

In this section we apply the generic derivations of the previous section to a simple binary system in a circular orbit. Consider two bodies with $m_1 > m_2$ on the relatively wide orbit so that all assumptions of the previous section are satisfied. We place the coordinate frame in the center of mass of the system and the orbit lies in the $x - y$ plane. We show a schematic set-up in Figure 2.3.

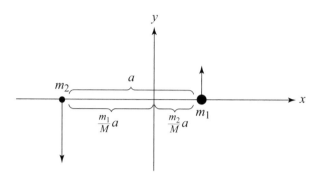

Figure 2.3: Sketch showing the binary system of two point masses $m_1 > m_2$ in a circular orbit around the common center of mass (at the origin of coordinate frame).

According to Kepler's law the angular orbital frequency is

$$\omega = \sqrt{\frac{M}{a^3}}, \tag{2.50}$$

where $M = m_1 + m_2$ is the total mass and a is the orbital separation. Assume that at $t = 0$ the bodies are along the x-axis, then equations of motion can be written as

$$x_1 = \frac{m_2}{M}a\cos\omega t, \quad y_1 = \frac{m_2}{M}a\sin\omega t \tag{2.51}$$

$$x_2 = -\frac{m_1}{M}a\cos\omega t, \quad y_2 = -\frac{m_1}{M}a\sin\omega t. \tag{2.52}$$

We neglect the size of the bodies compared to the orbital separation, so we can treat each object as a point mass:

$$I_{jk} = \int T^{00}x_j x_k d^3x = \int \left[\delta\left(\vec{x} - \vec{x}_1\right)m_1 + \delta\left(\vec{x} - \vec{x}_2\right)m_2\right]x_j x_k d^3x$$
$$= m_1 x_1^j x_1^k + m_2 x_2^j x_2^k. \tag{2.53}$$

Using equations of motion and Kepler's law we obtain

$$\ddot{I}_{xx} = -\ddot{I}_{yy} = -2\mu\left(M\omega\right)^{2/3}\cos 2\omega t, \tag{2.54}$$

$$\ddot{I}_{xy} = \ddot{I}_{yx} = -2\mu\left(M\omega\right)^{2/3}\sin 2\omega t, \tag{2.55}$$

where we have introduce a reduced mass ratio $\mu = m_1 m_2/M$. Next, we will introduce a polarization basis, and the easiest choice is the unit vectors $\hat{e}_\theta$, $\hat{e}_\phi$ tangent to the coordinate lines in the spherical polar coordinates orthogonal to the direction of GW propagation, or, in other words, direction to the observer (GW detector). We also assume a particular case $\phi_{observer} = 0$: observer lies in $x - z$ plane, and we will generalize it a bit later, so the polarization basis vectors are

$$\hat{e}_\theta = \hat{e}_x \cos\theta - \hat{e}_z \sin\theta, \quad \hat{e}_\phi = \hat{e}_y. \tag{2.56}$$

The angle θ is a polar angle between normal to the orbital plane (or the orbital angular momentum which is chosen to be aligned with z-axis) and the direction to the observer. Since we use spherical polar coordinates, it is convenient to transform the obtained above tensor of inertia (the second moment of mass distribution) using the generic coordinate transformation rule $x_{\bar{k}} = \frac{\partial x^i}{\partial x^{\bar{k}}} x_i$ to the spherical coordinates:

$$I_{\theta\theta} = I_{xx} \cos^2\theta, \quad I_{\phi\phi} = I_{yy}, \quad I_{\theta\phi} = I_{xy} \cos\theta. \tag{2.57}$$

Now it is time to generalize this expression to an arbitrary azimuthal position of the observer, to do so we can rotate the line of sight (observer) around z-axis by some angle ϕ_0. We claim that it is equivalent to the rotating of the binary by $-\phi_0$ at initial moment of time $t = 0$, in other words changing the phase $\omega t \to \omega t - \phi_0$. Indeed, in this particular case, those two angles are completely degenerate. Note that this is not true, in general; in particular, if we have spinning bodies with an arbitrary orientation of spins.

Let us also say a few words about the angle θ. It is the angle between the orbital angular momentum of the binary and the direction of the wave propagation (direction to the observer from the source frame). Another angle is often used instead in the literature; the angle between the line of sight (direction to the source from the detector's frame) and the orbital angular momentum: $\iota = \pi - \theta$. In this book, we use both notations, depending if we work in the source or in the detector based frame. Finally, the strain of GW emitted by a binary system is given as

$$h_+ = h_{\theta\theta} = -2\left(1 + \cos^2\theta\right) \frac{\mu}{R}(M\omega)^{2/3} \cos\left[2\omega(t - R) - \phi_0\right] \tag{2.58}$$

$$h_\times = h_{\theta\phi} = -4\cos\theta \frac{\mu}{R}(M\omega)^{2/3} \sin\left[2\omega(t - R) - \phi_0\right]. \tag{2.59}$$

Let us spend few minutes on these expressions.

The amplitude of GW is scaled as $\eta M/R$, where $\eta = m_1 m_2/M^2$ is a dimensionless symmetric mass ratio, so the amplitude is inverse proportional to the luminosity distance to the source, and scales linearly with the total mass of the system. The factor η tells us that the radiation is stronger from the system with comparable masses, in the next section we will consider the case where the masses are highly asymmetric.

Coming back to the inclination angle θ, we see that the amplitude is maximum if we observe the binary "face-on/off" (which corresponds to $\theta = 0, \pi$), and the source appears weaker if it is seen "edge-on" ($\theta = \pi/2$). This is already observed selection bias; we will see predominantly face-on/off sources as they appear brighter.

We can also notice that the dominant gravitational radiation is at twice orbital frequency.

So far we neglected the back-action of GWs on the emitting system. The GWs carry away energy and angular momentum:

$$Lum = -\frac{dE^{GW}}{dt} = \frac{1}{5}\left\langle \dddot{\mathcal{M}}_{ij}\dddot{\mathcal{M}}_{ij} \right\rangle = \frac{32}{5}\eta^2(M\omega)^{10/3}. \tag{2.60}$$

This energy is extracted from the total energy of the binary system:

$$E^{tot} = \frac{m_1m_2}{a} + \frac{m_1(\omega r_1)^2}{2} + \frac{m_2(\omega r_2)^2}{2} = -\frac{m_1m_2}{2a}. \tag{2.61}$$

The orbit shrinks as a result of the energy extraction:

$$\dot{E}^{tot} = \frac{m_1m_2}{2a^2}\dot{a} = -\frac{32}{5}\eta^2(M\omega)^{10/3}. \tag{2.62}$$

Using Kepler's law we can get the differential equation for the orbital separation (or for the orbital frequency) which can be easily integrated:

$$a = \left[\frac{256}{5}\eta M^3\,(t_c - t)\right]^{1/4}, \tag{2.63}$$

$$\pi f = \left[\frac{256}{5}\eta M^{5/3}\,(t_c - t)\right]^{-3/8}, \tag{2.64}$$

where t_c is a constant of integrations such that $a = 0$ at $t = t_c$ and it is called the *coalescence time*. The definition of coalescence time is not unique; often it is associated with the time when the GW frequency becomes infinite, however it is an artifact of our approximation. Basically, it tells us that our assumptions break down, in particular, the orbital velocity becomes comparable to the speed of light and we also cannot treat anymore bodies as point masses. In the next section, we give semi-analytic and numerical schemes for solving Einstein equations for the two-body problem, with the right approach the GW frequency remains finite, and the coalescence time can be associated with the maximum of GW strain amplitude. Note that the combination of masses entering the frequency evolution is $\eta^{3/5} M \equiv M_c$, which is called a *chirp mass*.

We can integrate the orbital frequency to get evolution of the orbital phase:

$$\phi_{orb}^{(N)} = \int 2\pi f(t)\,dt = -2\left[\frac{1}{5M_c}\,(t_c - t)\right]^{5/8} + \phi_c, \tag{2.65}$$

where ϕ_c is an integration constant which is related to $\phi_0 = \phi(t = 0)$. The label (N) reminds us that we are working in the leading (Newtonian) order. This solution can be plugged back into the equation of motion for two bodies and we update the solution of Einstein equations iteratively. This scheme is called *Post-Newtonian approximation*.

As we will show later, detection of the GW signal and estimation parameters of the emitting system relies on the accurate tracking of the GW phase, which, at leading order, depends on M_c, so we already can expect that the chirp mass is the parameter which we can determine the best. Let us also give the rate of change of the GW frequency (which is at leading order $f^{GW} = 2f$):

$$\frac{df^{GW}}{dt} = \frac{96}{5\pi} M_c^{5/3} (\pi f_{GW})^{11/3}. \tag{2.66}$$

Here we want to emphasize its strong dependence on the instantaneous GW frequency; the signal is almost monochromatic at low frequencies (wide separation) and it changes very rapidly as two bodies approach each other. Note that the amplitude of GW signal grows as $f^{2/3}$, so, as bodies spiral toward each other, the signal becomes stronger and sweeps to a higher frequency, that is why the GW signal is also called *chirp*. For the case where one mass is much larger than other, $M_c^{5/3} \approx (m_2/m_1) m_1^{5/3}$, the evolution could be slow due to smallness of m_2/m_1 even for the relativistic orbit.

Finally, we look at the time it takes for a binary to merge by inverting the Eq. (2.64):

$$\Delta t = \frac{5}{256 M^{5/3} \eta} (\pi f_{GW})^{-8/3}. \tag{2.67}$$

This is the time left to coalescence from the instantaneous GW frequency f_{GW}. Let us consider how long the GW signal lasts for a different type of binaries. We start with a binary neutron star system; we choose the total mass of binary to be $M = 3\,M_\odot$ (typical for such a system) and the decent sensitivity starts with 30 Hz. Let us recall that we work in the geometrical units where $M_\odot \approx 5 \times 10^{-6}$ sec, substituting those numbers we get $\delta t = 11.7\,\sec/\eta$; the neutron stars are approximately of the same mass so the duration of the GW signal in the LIGO/Virgo band is less than a minute. Later on, we will also consider a future space-based GW observatory LISA, which will observe coalescence of massive BH binaries with $M \sim 10^5 - 10^8\,M_\odot$. Let us take $M = 10^6\,M_\odot$, the sensitivity of LISA starts roughly at 10^{-4} Hz, substituting those numbers we get $\delta t \approx 34\,\mathrm{days}/\eta$ which is, for the equal mass binary, roughly 4 months. For the extreme mass ratio system (few solar mass object spiralling toward the massive BH), this time scale is several orders of magnitude higher.

Last, we want to give an expression for the GW strain in the frequency domain:

$$\tilde{h}_{+,\times}(f) = \int h_{+,\times}(t) e^{-2i\pi f t} \, dt.$$

We can get an analytic expression for the adiabatic inspiral where the amplitude is slowly changing (compared to the orbital time scale). The GW strain in the time domain is of the form $h(t) \sim A \cos(\Phi(t))$, substituting this expression in the Fourier integral, we see that we can use the stationary phase approximation: the main contribution comes from the non-oscillatory part of the integrand which corresponds to the stationary point of $\Phi(t) - 2\pi f t$. Decomposing the integrand around the stationary point (up to second order), we can perform integration analytically and get

$$\tilde{h}_+(f) = \sqrt{\frac{5}{6}\frac{1}{4\pi^{2/3}}}\frac{M_c^{5/6}}{R}f^{-7/6}\left(1 + \cos^2\theta\right)e^{i\tilde{\Psi}(f)},\tag{2.68}$$

$$\tilde{h}_\times(f) = i\sqrt{\frac{5}{6}\frac{1}{4\pi^{2/3}}}\frac{M_c^{5/6}}{R}f^{-7/6}2\cos\theta e^{i\tilde{\Psi}(f)}.\tag{2.69}$$

The phase of GW signal (leading order) in the frequency domain is given as

$$\tilde{\Psi}(f) = 2\pi f t_c - \phi_0 - \frac{\pi}{4} + \frac{3}{4}\left(8\pi M_c f\right)^{-5/3}.\tag{2.70}$$

Note that the amplitude of the GW strain in the frequency domain scales as $f^{-7/6}$, which is higher at low frequencies (slow inspiral). This happens because the binary spends many orbital cycles before it changes the frequency under the gravitational radiation reaction.

2.4 FROM ANALYTICAL RELATIVITY TO NUMERICAL RELATIVITY

In this section we give a summary of analytic methods for modeling the GW signal from a compact binary system. We also touch a full numerical solution (coming from the numerical relativity, NR) of a two-body problem in the context of validating analytical (or semi-analytical) models. In what follows, we will not give any mathematical proofs and refer to the existing literature; instead, we outline the main assumptions and methods.

2.4.1 POST-NEWTONIAN FORMALISM

In the previous section, we have used two main assumptions: (i) weak internal gravity of an isolated system and (ii) slow motion (as compared to the speed of light). It turns out that the obtained leading order solution does not require weak internal gravity (see, e.g., [100]). In the Post-Newtonian (PN) formalism we solve the Einstein equation by iterations expanding in small parameters v (remember that we have chosen $c = 1$, so it is expansion in v/c). We have already derived the quadrupole expression (2.39) which corresponds to the Newtonian limit $1/c \to 0$ [60, 91], where we have used only Newtonian laws of gravity. However, we have already mentioned the nonlinear terms denoted as $t^{\mu\nu}$, which we have completely neglected; now it is time to restore them. We have also mentioned that other moments should contribute to the GW, moreover we have shown that their contribution is of higher PN order (2.43).

We start again with the harmonic coordinates, which more general form is given as $\hat{h}^{\alpha\nu}{}_{,\nu} = 0$, where $\hat{h}^{\alpha\beta} = \sqrt{-g}g^{\alpha\beta} - \eta^{\alpha\beta}$, and the field equations could be written as

$$\Box\hat{h}^{\alpha\beta} = 16\pi\tau^{\alpha\beta},\tag{2.71}$$

where

$$\tau^{\alpha\beta} = |g|T^{\alpha\beta} + \frac{1}{16\pi}t^{\alpha\beta}, \tag{2.72}$$

where $t^{\alpha\beta}$ is Landau–Lifschitz pseudo-tensor, and the box operator is the flat-spacetime wave operator. Note that this is an exact form of the Einstein equations written in a convenient form using the harmonic coordinates. All nonlinear terms $\sim \hat{h}^2, \hat{h}^3, \dots$ are contained in $t^{\mu\nu}$ and give important contributions to the GW signal in consecutive iterations. Formally, the PN scheme is valid only in the near zone $r \ll \lambda^{GW}$ and it is formally divergent at infinity where we actually measure the GWs.

In further description of PN formalism we follow closely a very nice and complete review by L. Blanchet [32]. Outside the isolating system the gravitational field satisfies a vacuum Einsteins equation and can be solved by iterations using post-Minkowskian expansion in the small parameter G (or in other words, the post-linear expansion in small amplitude $\hat{h}$), which implies only the weak self-gravitation of the system. The post-Minkowskian expansion is perfectly valid in the far zone but divergent as $r \to 0$. It is convenient to expand the post-Minkowsian solution in the set of symmetric and tracefree multipole moments [135], those are mass and current multipole moments which we have briefly mentioned in the previous section. The multipolar post-Minkowsian solution can be extended to the near zone of the source where it can be decomposed in PN series v/c. It can be shown that the PN (interior with respect to the source solution) and post-Minkowskian (exterior solution) have an overlapping region of validity $L < r < \lambda$ where PN expanded post-Minkowsian multipole moments can be associated with the multipole moments of the source. This scheme is called matched asymptotic expansion, and was introduced by K. Thorne and improved and developed by Damour–Blanchet–Iyer. A similar (but somewhat different in details) approach was developed by Will and Wiseman [151].

The nonlinearity of the field equations plays an important role. At higher order iterations, we have a coupling of multipoles; one interesting coupling is the interaction of the mass M of the source with its quadrupole moment I_{ij} which appears at the relative 1.5 PN order $((v/c)^3)$. The way the field Equations (2.71) are written gives a wrong impression that GWs propagate along the null cone of the flat spacetime, which is not true and it is an artifact of moving the nonlinear terms to the right-hand side. The spacetime around the source is not flat and, as a result, we have a scattering of the GWs (quadrupolar wave in this case) off the Schwarzschild curvature generated by the system with the mass M; the process of scattering is continuous and it manifests itself as an integral over the entire past history of the source, although the main contribution comes only from the recent past. Such contribution is called *tail*; at higher PN orders (3 PN) we also have quadratic tail or "tail of tail" (coming from $M^2 \times I_{ij}$). We also have nonlinear memory term at 2.5 PN order which manifests that the quadrupole wave (obtained at leading order) becomes the source itself ($I_{ij} \times I_{ij}$ interaction).

For widely separated systems, $a_{orb} \gg r_1, r_2$ (where a_{orb} is orbital separation and $r_{1,2}$ is a characteristic size of each body) we can neglect completely the finite size of each body and treat them as point masses. Mathematically, it corresponds to use δ-functions, as we have done in the previous section. Such a treatment simplifies the calculations but comes with the price that most of the integrals become divergent at the location of the point masses, and this requires "self-field" regularization. In particular, in the PN approach two types of regularization are used: (i) Hadamard (*finite part*) regularization scheme which is convenient in practice but works up to 3 PN order; and (ii) dimensional regularization [33]. We again refer to [32] for a detailed description. However, one can avoid using the point masses and treat the bodies as two fluid balls of finite dimension [75]. This approach shows equivalent results for the equation of motion at 2 PN order. Another important approach was undertaken by Futamase and collaborators [65, 84]; they used scaling arguments and transformed the problem into integration over the surface surrounding bodies which can be seen as large in appropriately introduced "body coordinates" and arbitrary small in the "exterior coordinates." This approach is similar in spirit to [61] and it also can be seen as a matched asymptotic performed on the surface. This method was used at 3 PN order and shows equivalent results to the dimensional regularization.

At each PN order one needs to use a balance equation

$$\frac{dE}{dt} = -Lum \tag{2.73}$$

(similar to what we have done in the previous section) to update equations of motion. Note that the radiation-reaction force appears in the equation of motion at 2.5 PN order. We consider here only circular binaries so we use only the balance equation for the energy of the system.

We should also mention the finite size effect. In reality, the bodies in the binary system have finite size, internal structure, and strong gravity; it turns out that the effect of the internal structure appears only at 5 PN order $((v/c)^{10})$, however, it can play an important role in the binaries containing neutron stars (NSs), such as NS-NS and NS-BH binaries.

We want to emphasize in particular the Hamiltonian approach for deriving equations of motion [18, 86] as it has proven to be especially powerful in treating spinning binary systems. The Hamiltonian of a two-point mass system could be obtained within a canonical ADM [18] formalism in the form of PN expansion written in the center-of-mass frame:

$$H\left(\vec{X}, \vec{p}\right) = Mc^2 + H_N\left(\vec{X}, \vec{p}\right) + \sum_k \frac{1}{c^{2k}} H_{kPN}\left(\vec{X}, \vec{p}\right), \tag{2.74}$$

where Newtonian part is

$$H_N(\vec{X}, \vec{p}) = \frac{\vec{p}^2}{2\mu} - \frac{M\mu}{r} \tag{2.75}$$

and we have used relative separation vector $\vec{X} = \vec{X}_1 - \vec{X}_2$, and momentum $\vec{p} = \vec{p}_1 = -\vec{p}_2$, the subscript 1 corresponds to heavier BH. The Hamiltonian is currently known up to 4 PN order [117].

Finally, the GW polarizations can be written as PN series (we use here again $G = c = 1$)

$$h_{+,\times} = \frac{2\mu x}{R} \left\{ h_{+,\times}^{(0)} + x^{1/2} h_{+,\times}^{(1/2)} + x h_{+,\times}^{(1)} + x^{3/2} h_{+,\times}^{(3/2)} + x^2 h_{+,\times}^{(2)} + x^{5/2} h_{+,\times}^{(5/2)} + O(x^3) \right\},$$

(2.76)

where $x = (M\omega)^{2/3} = O(v^2)$ and we give explicitly only the first two leading order terms

$$h_+^{(0)} = -(1 + \cos^2 \theta) \cos 2\Psi \tag{2.77}$$

$$h_+^{(1/2)} = -\frac{\sin \theta}{8} \frac{\delta m}{M} \left[(5 + \cos^2 \theta) \cos \Psi - 9(1 + \cos^2 \theta) \cos 3\Psi \right] \tag{2.78}$$

$$h_\times^{(0)} = -2 \cos \theta \sin 2\Psi \tag{2.79}$$

$$h_\times^{(1/2)} = -\frac{3}{4} \sin \theta \cos \theta \frac{\delta m}{M} [\sin \Psi - 3 \sin 3\Psi], \tag{2.80}$$

where

$$\Psi = \phi - 2M_{ADM}\omega \ln \left(\frac{\omega}{\omega_0} \right) \tag{2.81}$$

and $M_{ADM} = M(1 - \eta x/2)$ is the ADM mass (in next-to-leading order). One can recognize already derived leading order expressions (2.59). Note also that sub-leading order terms have first and third harmonics of the orbital frequency and those terms disappear for exactly even mass binary $\delta m = m_1 - m_2 = 0$. The orbital phase is also represented as a Taylor series in PN parameter x:

$$\phi = \phi_N \left[1 + \left(\frac{3715}{1008} + \frac{55}{12} \eta \right) x - 10\pi x^{3/2} + \ldots \right] \tag{2.82}$$

$$\phi_N = -\frac{x^{-5/2}}{32\eta}. \tag{2.83}$$

Higher-order PN terms for GW polarization and for the orbital phase can be found in [32].

Here we have used x as a parameter for PN expansion; one could also use $\gamma = M/r \sim O((v/c)^2)$ parameter which is related to x also via Taylor expansion. Representing the phase in γ and truncating the series at appropriate PN order gives us a mathematically equivalent expression. Note that we also need to express γ and/or x as a function of time which is yet another PN Taylor series. All these different truncated representations of a GW signal agree for mildly relativistic parts of the binary evolution (slow motion, low frequency) but they diverge as we are coming closer to the merger. The binary consisting of two neutron stars has low total mass, so the merger happens at a high frequency where the sensitivity of advanced LIGO has already degraded (due to rising shot noise) and therefore any representation gives waveforms which are similar to sufficiently good precision. The binary black hole system could have significantly high mass (for example GW150914 has a total mass $M \sim 65 \, M_\odot$) and they merge right in the most sensitive part of the LIGO frequency band. The observed end of inspiral enters the ultra-relativistic regime where the accuracy of the truncated PN series is not sufficient, and different approximations diverge. Different approximants[1] use different parametrization of an exact signal and being truncated at a given PN order capture slightly different parts of an exact solution; the difference, being small, is amplified by integration (accumulated over integration) and we also enter the part of the evolution where some assumptions made for the PN scheme break down. One more important approximant could be obtained using the Fourier transformed GW signal (PN extension of the stationary phase approximation given in the previous section).

In addition to the different parametrization, one can use other re-summation techniques, the most useful of which was proven to be the Padé re-summation, which is often used if there is a pole which makes a series expansion poorly convergent (might be even divergent) as we push it to the limits of validity. The Padé re-summation operates with fractions:

$$P_{m,n}(x) = \frac{N_m(x)}{D_n(x)}, \tag{2.84}$$

where numerator $N_m(x)$ and denominator $D_n(x)$ are polynomials in x of order m and n correspondingly, and the Taylor expansion of $P_{m,n}(x)$ coincides with a desired (Taylor expanded) function at a given truncated order. The Padé expansion is applied to the total energy (actually to its derivative) and to the GW flux which enter the balance equation and, therefore, the orbital evolution. To be more precise, the Padé re-summation is applied to the auxiliary energy and flux-type functions motivated by the test mass limit ($\eta \to 0$) and factoring out the logarithmic terms (in flux), the reader can find more details in [38, 50] and references therein. This re-summation improves the convergence of the original Taylor series.

Here we have considered only the non-spinning binary system on the circular orbit. We refer the reader to [19] where an explicit PN expression for GW polarizations of two spinning (with arbitrary spin orientation) is given. The PN description of mildly eccentric systems can be found in [48, 133].

2.4.2   EXTREME MASS RATIO INSPIRALS (EMRIs)

In this section we consider a specific binary system with very uneven mass ratio, namely with $m_1 \gg m_2$ so that $M \approx m_1$ and $m_2 \equiv m$, $m/M \sim 10^{-4} - 10^{-7}$. We believe that such systems could be formed in the galactic nuclei which host massive black holes (MBHs) $M \sim 10^5 - 10^7\,M_\odot$ and a small compact object (CO) (solar mass black hole, neutron star or white dwarf) could be thrown toward (as a result of N-body interaction) and captured by an MBH. The characteristic feature of such a system is a very slow inspiral; according to Eq. (2.67) the inspiral time scales inversely with the $M^{5/3}\eta$, where η is of order of mass ratio. This implies that the CO can spend a long time in the close vicinity of an MBH, where the PN expansion has limits of its validity. Instead of going to very high PN orders we can try to solve the Einstein equations

---

[1]Approximants are the GW models which use different PN parametrization and truncated at different PN level.

using another small parameter: mass ratio m/M. In the limit of a test mass $m/M \to 0$, the CO moves on the geodesic orbit. The geodesic equation of motion could be obtained starting from the "super-Hamiltonian"

$$\mathcal{H}(x^\alpha, p_\alpha) = \frac{1}{2} g^{\mu\nu} p_\mu p_\nu = -\frac{1}{2} m^2, \tag{2.85}$$

where $p^\alpha = dx^\alpha/d\tau$ is 4-momentum of a test mass and we will use the Kerr metric in the Boyer–Lindquist coordinates which allow full separation of variables. We refer the reader to [72] for more details on the EMRIs motion and waveforms and outline here only the main aspects which we will need later in the book. The Hamiltonian can be written in terms of the action-angles variables:

$$J_i = \oint p_\alpha dx^\alpha = \begin{cases} J_r &=& 2\int_{r_p}^{r_a} dr\, \sqrt{R}/\Delta \\[2mm] J_\theta &=& 2\int_{\theta_{\min}}^{\theta_{\max}} d\theta\, \sqrt{\Theta} \\[2mm] J_\phi &=& 2\pi L_z, \end{cases} \tag{2.86}$$

where index α here does not assume summation, Δ was introduced in Eq. (1.8), r_a, r_p are the apoaps and periaps (furtherest and closest approach points of the orbit) and $\theta_{\min}, \theta_{\max}$ are polar angles limiting the orbital precession due to spin-orbital coupling. The auxiliary functions are defined as

$$R = T^2 - \Delta\left[m^2 r^2 + (L_z - aE)^2 + Q\right], \quad T = E(r^2 + a^2) - L_z a \tag{2.87}$$

$$\Theta = Q - \cos^2\theta\left[a^2\left(m^2 - E^2\right) + \frac{L_z^2}{\sin^2\theta}\right], \tag{2.88}$$

where one can see explicit dependance on the first integrals E, L_z, Q. The geodesic orbit can be seen as a three-periodic motion in r, θ, and ϕ coordinates, moreover, the three corresponding frequencies are in general non-commensurate. The test mass goes around MBH (ϕ-motion), due to the presence of eccentricity it moves between r_p and r_a in the radial direction, and the ellipse is precessing (relativistic precession similar to what is observed in the solar system for the orbit of Mercury) and, if the orbit is not equatorial, there is an orbital precession (precession of the orbital angular momentum around the spin of an MBH) which corresponds to the θ-motion. Three orbital frequencies are given by

$$\omega_i = \frac{\partial \mathcal{H}}{\partial J_i}; \tag{2.89}$$

one more "frequency" is associated with the time-coordinate:

$$\omega_t = -\frac{\partial \mathcal{H}}{\partial E}, \tag{2.90}$$

where E is the energy of a test mass (one of the constants of motion). This frequency appears as a result of parametrization of the trajectory by proper time $t = t(\tau)$ which we use as an affine parameter, and the physical (measurable at infinity by an observer) frequencies are

$$M\Omega_i = \omega_i/\omega_t. \tag{2.91}$$

The geodesic motion is a non-physical approximation $m/M \to 0$; the motion of a CO object around MBH is accompanied by gravitational radiation, and, as a result, the orbit shrinks and circularizes. We have already mentioned that the radiation reaction scales as M/m which is, in the case of EMRIs, $T_{RR} \gg T_{orb}$ so the motion of a CO around MBH can be seen as a slow drift from one geodesic to another in an adiabatic way so we can use the osculating elements approach [36, 66, 73, 110].

   We assume that at the zero order in a mass ratio the CO moves on the geodesic and use it for computing the GW flux. The GW strain and fluxes can be computed using Teukolsky formalism [134] which encapsulates all gravitational radiative degrees of freedom in a single "master" wave equation for the Weyl scalars, Ψ_0 and Ψ_4. The rate of change in energy (dE/dt), angular momentum (dL_z/dt), averaged over several orbits are evaluated from the gravitational wave field. The "Carter flux" (the rate of change the Carter constant Q) cannot be inferred from the GW radiation at infinity alone. As shown in [87], we need a local (with respect to CO) expression for the self-force (we will discuss the self-force shortly). There is, nevertheless, PN prescription for computing the Carter flux (dQ/dt) averaged over several orbital periods. It is given in [68] and it explicitly depends on some (local instantaneous) orbital parameters. One could use the balance equation and update the motion of a CO but it wouldn't be complete. Here we need to introduce a concept of a *self-force* which we actually already use in the PN description. The self-force can be seen to arise as a result of an interaction of the self-field of the CO with the non-flat background geometry, which causes the lines of force to be bent and act back on the CO. The self-force can be conventionally split into two parts: time non-symmetric (dissipative) and time-symmetric (conservative). The former part causes the inspiral and dominates while the latter part can be eliminated by a redefinition of the orbital frequencies at each instance, which means that it is effectively of second order in the mass ratio. The adiabatic fluxes based on the Teukolsky waveforms take into account only the dissipative part of the self force, neglecting the conservative part, which defines the domain of its validity. We refer readers to the very extensive review on this subject given in [108]. Here we mention only one particular approach called the "effective source approach." The computation of the self-force is somewhat complicated as it often treats the CO as a delta function in the background spacetime, which requires mathematical apparatus for the regularization of some divergent integrals (similar to what is done in the PN theory). It is possible to subtract the singular part from the field equations (by computing a singular solution valid in the vicinity of the CO) and the resulting equations are manifestly regular and contain on the right-hand side a smooth effective source [142], which allows the field equations to be coupled to the equations of motion and integrated. This procedure can be written for a scalar field (representing a CO carrying a scalar charge and ignoring the gravitational part of the self-force) as [58]:

$$(\Phi^r)_{;\alpha}{}^{;\alpha} = S(x; z(\tau), u(\tau)) \tag{2.92}$$

$$\frac{Du^\alpha}{d\tau} = \frac{q}{m(\tau)} \left(g^{\alpha\beta} + u^\alpha u^\beta \right) \nabla_\beta \Phi^r \tag{2.93}$$

$$\frac{dm}{d\tau} = -q u^\beta \nabla_\beta \Phi^r, \tag{2.94}$$

where Φ^r is a scalar field, q is the scalar charge, S is an effective (smooth) source term, and Du^α is a covariant derivative of a 4-velocity of a CO. The leading order self-force on the Schwarzschild background was fully computed and the problem is almost solved for the Kerr background. Recently, there was big progress toward finding the second-order self-force; the method is fully developed and applied to the simplest case of a cirular orbit in Schwarzschild background [28].

As we have mentioned several times, the GW signal from EMRIs is long-lived (depending on the proximity of the source, it could be "visible" for the duration of whole data taking) and, in comparison to the comparable mass binaries, the signal-to-noise ratio grows roughly as a root square of the observation time. This implies that we need a sufficiently accurate template (model) to describe $10^5 - 10^6$ GW cycles. The GW signal from EMRIs could be written as a sum of harmonics of three slowly changing orbital frequencies described previously:

$$h(t) = \sum_{l,m,n} h_{lmn}(t) = \Re \left(\sum_{lmn} A_{lmn}(t) e^{(l\Phi_r(t) + m\Phi_\theta(t) + n\Phi_\phi(t))} \right), \tag{2.95}$$

where the phases Φ_r, Φ_θ, Φ_ϕ are obtained from the integration of the corresponding orbital frequencies. The amplitudes of each harmonic $A_{lmn}(t)$ are functions of the orbital parameters (eccentricity, inclination (angle between the spin and the orbital angular momentum), orbital separation) which also evolve as CO spirals toward the MBH. As a result, the dominant harmonics could change during the inspiral (for example due to orbital circularization). We refer to reviews [23, 28, 108, 149] describing a present status of EMRIs GW modeling.

2.4.3   EFFECTIVE-ONE-BODY APPROACH

In this section, we describe (without going too much into details) another method for computing GW signal, it is called *Effective One Body approach* (EOB) [37, 39, 47]. This method encapsulates the current knowledge from PN theory with parts from the test mass (better to say "self-force") calculations described in the EMRIs section. The basic idea is to map the full relativistic dynamics of the two-body problem (with arbitrary masses) to the problem of a test mass moving in the effective background. Let us start with the description of two non-spinning BHs (as we have done it in PN section) to introduce the main idea of this approach. EOB can be constructed using three ingredients: (i) description of the conservative dynamics of two-body system, (ii) dissipation due to emission of the GWs, and (iii) construction of the GW signal.

We start with the conservative dynamic, neglecting for a time being any backreaction on the system due to gravitational emission. The main idea is to map the motion of two, say, comparable mass BHs to the problem of a test mass m moving in the *deformed* Schwarzschild spacetime of a mass M. The deformation parameter is a symmetric mass ratio η, and as $\eta \to 0$ we recover the Schwarzschild metric. By construction, this approach reduces to the EMRI problem described in the previous section if $m_2/m_1 \ll 1$. Now we give some details on how to achieve such mapping. The mapping is done using Hamiltonians. We start with the PN Hamiltonian of two-body problem partially given in (2.74). In what follows, we will use Hamiltonian to third PN order and we refer the reader to [46] and [52] for the full form of the Hamiltonian given in the center-of-mass. Note that recently 4-PN order was computed as well [53], where it was shown that the Hamiltonian contains also the "non-local" in time term due to tail-transported temporal correlations, which make the equations of motions at time t depend on the state of the system in the past.

The Newtonian part of the Hamiltonian (2.75) already can be seen as a test mass μ (reduced mass) moving in spherical potential generated by a mass M (total mass); EOB, in this respect, is the relativistic generalization of this Newtonian interpretation. The "effective" dynamics is constructed as a motion of a test mass μ in an effective spacetime:

$$ds^2_{\text{eff}} = -A(r_{\text{eff}}, \eta)dt^2 + \frac{D(r_{\text{eff}}, \eta)}{A(r_{\text{eff}}, \eta)}dr^2_{\text{eff}} + r^2_{\text{eff}}d\Omega^2, \tag{2.96}$$

where we have explicitly introduced r_{eff} not to confuse it with the vector $\vec{r}$ used in the two-body Hamiltonian. The metric coefficients $A(r_{\text{eff}}, \eta)$, $D(r_{\text{eff}}, \eta)$ will be determined from the matching between two problems, however our desire to have Schwarzschild limit as $\eta \to 0$ already sets the constrains

$$A(r_{\text{eff}}, \eta = 0) = 1 - \frac{2M}{r_{\text{eff}}}, \quad D(r_{\text{eff}}, \eta = 0) = 1. \tag{2.97}$$

We can use those functions in the form of expansion in M/r_{eff} with coefficients depending on η, which can be seen as a deformation away from the Schwarzschild metric when we deal with the comparable masses. We demand that the dynamics of a test mass μ is described by

$$g^{\mu\nu}_{\text{eff}} p^{\text{eff}}_\mu p^{\text{eff}}_\nu + \mu^2 + Q\left(p^{\text{eff}}_\mu\right) = 0. \tag{2.98}$$

Here, $Q(p^{\text{eff}}_\mu)$ is *not* a Carter constant but a term which is at least quartic in p^{eff}_μ. Equation (2.98) can be seen as a generalization of (2.85); here we have used an effective metric and also introduced Q-term.

Both real and effective Hamiltonians can be presented in the action-variables (as we have done for the geodesic motion in the previous section; see Eq. (2.86)). The authors in [37], being inspired by quantum mechanics, have performed the matching using Delaunay Hamiltonians $\mathcal{E}(N, J_\phi)$ (with $N = h\hbar = J_r + J_\phi$, $J_\phi = l\hbar$ and as before $J_\phi = (1/2\pi)\oint p_\phi d\phi$, $J_r = (1/2\pi)\oint p_r dr$) and $\mathcal{E}_{\text{eff}}(N_{\text{eff}}, J^{\text{eff}}_\phi)$. In other words, the idea is to match the "energy levels of bound states" of the real two-body problem and a test mass moving in the effective spacetime:

$$\mathcal{E}_{\text{eff}}(N_{\text{eff}}, J_{\phi}^{\text{eff}}) = f[\mathcal{E}(N, J_{\phi})]. \tag{2.99}$$

It was suggested to look for a mapping function f in the form

$$\frac{\mathcal{E}_{\text{eff}} - m}{m} = \frac{\mathcal{E}^{\text{NR}}}{\mu}\left[1 + \alpha_1 \frac{\mathcal{E}^{\text{NR}}}{\mu} + \alpha_2 \left(\frac{\mathcal{E}^{\text{NR}}}{\mu}\right)^2 + \alpha_3 \left(\frac{\mathcal{E}^{\text{NR}}}{\mu}\right)^3 + \dots\right], \tag{2.100}$$

where $\mathcal{E}^{\text{NR}} = \mathcal{E} - M$ is a "non-relativistic energy." The metric coefficients are searched in the form of the Taylor expansion:

$$A(r) = 1 + a_1 \frac{M}{r} + a_2 \left(\frac{M}{r}\right)^2 + a_3 \left(\frac{M}{r}\right)^3 + \dots \tag{2.101}$$

$$D(r) = 1 + d_1 \frac{M}{r} + d_2 \left(\frac{M}{r}\right)^2 + d_3 \left(\frac{M}{r}\right)^3 + \dots \tag{2.102}$$

In order to map two Hamiltonians while keeping the action variables invariant we need to apply a canonical transformation using the generating function $\tilde{G}(r, p_{\text{eff}}) = r^i p_i^{\text{eff}} + G(r, p_{\text{eff}})$.

Let us explain the reasoning behind the quartic (last) term in Eq. (2.98). At each PN order we need to determine three unknown coefficients a_{n+1}, d_n, α_n, as we have mentioned before, the Newtonian limit requires $a_1 = -2$. The difference between the number of equations and the number of unknowns at nth PN order is $n - 2$. This was the main motivation behind introducing the "post-geodesic" term which form was inspired by terms with higher order derivatives in the effective action for the scalar field [51]. As discussed in [37], it is natural to associate the test mass $m = \mu$ and and the mass of the central object $M = m_1 + m_2$. The generalized effective Hamiltonian then can be written as

$$H_{\text{eff}}(\vec{r}_{\text{eff}}, \vec{p}_{\text{eff}})/\mu = \sqrt{A(r_{\text{eff}})\left[1 + \vec{p}_{\text{eff}}^2 + \left(\frac{A(r_{\text{eff}})}{D(r_{\text{eff}})} - 1\right)(\vec{n}_{\text{eff}} \cdot \vec{p}_{\text{eff}})^2 + \frac{Q_{3PN}(p_{\text{eff}})}{r_{\text{eff}}^2}\right]}, \tag{2.103}$$

where

$$Q_{3PN}(p_{\text{eff}}) = z_1 \left(\vec{p}_{\text{eff}}^2\right)^2 + z_2 \vec{p}_{\text{eff}}^2 (\vec{n}_{\text{eff}} \cdot \vec{p}_{\text{eff}})^2 + z_3 (\vec{n}_{\text{eff}} \cdot \vec{p}_{\text{eff}})^4 \tag{2.104}$$

and as before $\vec{n}_{\text{eff}} = \vec{r}_{\text{eff}}/r_{\text{eff}}$. This term adds three arbitrary constants z_1, z_2, z_3 at 3 PN order, and only one can be determined by matching—we choose $z_1 = z_2 = 0$. The matching between real and effective Hamiltonians is given by

$$H = M\sqrt{1 + 2\eta \frac{H_{\text{eff}} - \mu}{\mu}} \tag{2.105}$$

and the metric coefficients together with the quartic term are determined through the matching condition (up to 3 PN order which we consider here):

$$A(r) = 1 - \frac{2M}{r} + 2\eta \left(\frac{M}{r}\right)^3 + \eta \left(\frac{94}{3} - \frac{41}{32}\pi^2\right) \left(\frac{M}{r}\right)^4 \qquad (2.106)$$

$$D(r) = 1 - 6\eta \left(\frac{M}{r}\right)^2 + 2\eta(3\eta - 26) \left(\frac{M}{r}\right)^3 \qquad (2.107)$$

$$Q = 2\eta(4 - 3\eta) \left(\frac{M}{r}\right)^2 \frac{p_r^4}{\mu^2}. \qquad (2.108)$$

The extended version of EOB, which also includes the spins of BHs, is given in the series of papers [27, 55, 103, 128, 129] and will not be discussed here. We also refer to a review [47] for a more detailed description of EOB formalism.

So far, we have determined the dynamics; moreover we have neglected the dissipation due to gravitational radiation. Let us recall that we consider here only circular binaries, so it is sufficient to include radiation reaction force in the Hamiltonian equation for p_ϕ. Since we deal with the non-spinning BHs, it is convenient to work in spherical polar coordinates r, ϕ, p_r, p_ϕ, where equations of motion are

$$\frac{d\phi}{dt} = \Omega = \frac{\partial H}{\partial p_\phi} \qquad (2.109)$$

$$\frac{dr}{dt} = \frac{\partial H}{\partial p_r} \qquad (2.110)$$

$$\frac{dp_\phi}{dt} = F_\phi \qquad (2.111)$$

$$\frac{dp_r}{dt} = -\frac{\partial H}{\partial r}. \qquad (2.112)$$

The force is closely related to the balance equation, which in a Hamiltonian approach can be found as [39]

$$\left(\frac{dH}{dt}\right) \approx \Omega F_\phi, \qquad (2.113)$$

where we have neglected the terms of order $O(\dot{r}^2)$ (quasi-circular orbit) and we can relate it to the gravitational luminosity *Lum*:

$$F_\phi = -\frac{1}{\Omega} \left(\frac{2}{16\pi} \sum_\ell \sum_{m=-\ell}^{\ell} (m\Omega)^2 |Rh_{\ell m}|^2\right). \qquad (2.114)$$

We can use a particular form of the resumed waveform suggested in [49] and we have decomposed the GW signal in the spin weighted (where the spin is -2) spherical harmonics:

$$h_{\ell m} = h_{\ell m}^N \hat{h}_{\ell m}^{PN}, \tag{2.115}$$

$$h_{\ell m}^N = \frac{M\eta}{R} n_{lm}^{(\epsilon)} c_{\ell+\epsilon}(\eta) x^{(\ell+\epsilon)/2} Y^{\ell-\epsilon,-m}\left(\frac{\pi}{2}, \phi(t)\right), \tag{2.116}$$

where $x = (M\Omega)^{2/3}$, R is the luminosity distance to the source, ϵ denotes the parity and it is $\epsilon = 0$ for the mass-generated and $\epsilon = 1$ for the current-generated multipoles, and $Y^{\ell m}(\theta, \phi)$ are the scalar spherical harmonics. We have also introduced the following auxiliary functions:

$$n_{\ell m}^{(0)} = (im)^\ell \frac{8\pi}{(2\ell+1)!!} \sqrt{\frac{(\ell+1)(\ell+2)}{\ell(\ell-1)}} \tag{2.117}$$

$$n_{\ell m}^{(1)} = -(im)^\ell \frac{16\pi i}{(2\ell+1)!!} \sqrt{\frac{(2\ell+1)(\ell+2)(\ell^2-m^2)}{(2\ell-1)(\ell+1)\ell(\ell-1)}} \tag{2.118}$$

$$c_{\ell+\epsilon}(\eta) = \left(\frac{m_2}{M}\right)^{\ell+\epsilon-1} + (-1)^m \left(\frac{m_1}{M}\right)^{\ell+\epsilon-1}. \tag{2.119}$$

The factorized PN part $\hat{h}_{\ell m}^{PN} = 1 + h_1 x + h_{1.5} x^{3/2} + \dots$ is presented as a product of four terms given explicitly in [49]; it is lengthy and we will not give it here.

So far, we have explained how to compute the dynamics of the binary system by mapping the real two-body problem to the effective dynamics of μ moving in the deformed Schwarzschild (or Kerr for spinning BHs) spacetime. We have added a dissipation and explained how to compute the waveform in the resummed form and expanded in the spherical harmonics. Two BHs will eventually come into contact with each other (merger) where the equations describing Hamiltonian dynamics are not applicable anymore. We push the dynamics to the point where the orbital frequency $\Omega(t)$ has a maximum, interestingly it happens near the *light-ring*.[2] After that we attach a *ring-down* part of the signal. The ring-down GW signal is obtained independently of the EOB approach by solving the linearized Einstein equations for an excited single BH. This GW signal is described by a superposition of exponentially damped eigen frequencies of a BH. Those frequencies and damping times (often referred to as complex quasi-normal-mode frequencies) are functions of mass and spin of a BH, however the amplitude of each mode depends on the way BH is excited. We use several quasi-normal-modes (usually eight modes) to describe the ring-down part of the signal which we smoothly attach to the previously described inspiral-merger part of the GW signal:

$$\left(\frac{R}{M}\right) h_{\ell m}^{RD}(t > t_{\text{match}}) = \sum_{m',n} A_{\ell,m',n} e^{-\sigma_{\ell,m',n}(t-t_{\text{match}})}, \tag{2.120}$$

---

[2]The last stable circular orbit around a black hole for the massless particle defines the light-ring.

where $\sigma_{\ell,m',n} = 1/\tau_{\ell,m',n} + i\omega_{\ell,m',n}$ are the quasi-normal frequencies and n is the "overtone index." Note that m' is not necessarily the same as m on the left-hand side. The matching time t_{match} is determined as a maximum of the orbital frequency (as already mentioned) or as the maximum of the amplitude of a harmonic (whatever happens first). The unknown amplitudes of each overtone are found by demanding the smoothness of the inspiral-merger waveform and its derivatives over some (usually small) interval prior to the t_{match}; for details, see for example [129]. The schematically this procedure is shown in Figure 2.4. In that figure we show the inspiral-plunge trajectory computed from EOB dynamics (in green) in the left plot. In the right part, we plot the orbital frequency (in units $2M$) in the lower panel and the full $\ell = 2, m = 2$ mode in the top panel. The black vertical dashed line corresponds to the t_{match}. The yellow part of the waveform is an attached h_{22}^{RD}. The middle panel gives the frequency of h_{22} in units of M. The red vertical dashed line corresponds to the last stable circular orbit for a test mass in the effective spacetime. Here we gave only a basis of the EOB approach and we invite the readers to the nice review by Damour [47] for more details and references for the further reading.

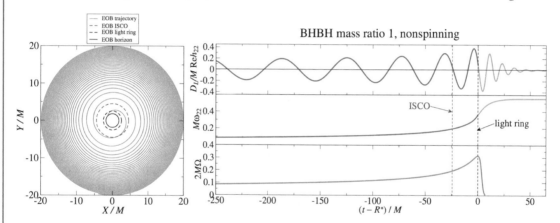

Figure 2.4: Left plot: The inspiral-plunge trajectory. Right plot: the EOB waveform with attached ring-down part and GW instantaneous frequency evolution (ω_{22}). This figure is courtesy of A. Taracchini.

2.4.4   NUMERICAL RELATIVITY

Numerical relativity (NR) is the branch of relativistic astrophysics which deals with the strong gravity where the Newtonian gravity is not sufficient. NR solves the Einstein equations numerically without any approximations, and therefore exactly within the numerical accuracy. In particular, here we are interested in solving the two-body problem (more precisely two BHs) numerically. After decades of struggle, several groups have managed to evolve a binary system to a merger. The longest available waveform is computed by SXS collaboration (Cornell-Caltech-CITA) and consists of 350 GW cycles (non-spinning BHs with mass ratio 1:7) [126]. Hundreds of waveforms have been computed to date which differ in the implementation and, as a result, in the numerical accuracy, they span rather a moderate mass ratio (up to $m_1/m_2 \sim 10 - 15$) and rather moderate spins (only a few dozen recent simulations have the spins above $0.6m_{1,2}^2$). Nevertheless, the NR simulations are extremely valuable to test the analytical approximations in order to understand the limits of their validity and to improve their precision.

In the previous section, we described the EOB approach which could be used to produce the inspiral-merger-ringdown waveform. The comparison of EOB and NR waveforms have confirmed that the idea behind EOB works quite well, but the accuracy is not always sufficient for the parameter estimation and, even sometimes, for the detection purposes (we will discuss it in the next chapter). The reason for lack of accuracy is twofold. (i) As two bodies approach each other the contribution of high PN order terms, some of which are unknown, becomes significant (still small but not negligible). We can foresee the functional form of missing terms and determine the coefficients through fitting to NR waveforms. Close to the plunge the adiabatic quasi-circular motion is not, strictly speaking, valid. To overcome this, the *non quasi-circular coefficients* are introduced in the factorized waveform and these coefficients are determined from the comparison to NR waveforms around the merger. (ii) We attach the ring-down part of the signal to the inspiral-merger, however the ring-down and the corresponding quasi-normal modes are obtained within the linear perturbation theory. It is hard to call two merging BHs a "linear perturbation," and, as a result, the ring-down part of the signal requires some artificial changes in some parts of the parameter space. Another (almost) phenomenological prescription for the ring-down signal was suggested [54], where the authors factor out the less damped quasi-normal mode and describe the rest of the waveform by an analytic model with the parameters determined from the fit to the NR signals. The extension of the EOB model using numerical data is called a *calibration* of the EOB or EOBNR model.

Another important class of the inspiral-merger-ringdown (IMR) waveforms was developed exclusively thanks to the numerical relativity. These are so-called *phenomenological IMR models* [12, 77, 89, 115]. The waveforms are generated in the frequency domain by (continuously) stitching together the inspiral part (known analytically from the PN approximation, see Section 2.4.1) with the merger part and ringdown. The inspiral part is presented by a high-order PN expansion of the stationary phase approximation introduced at the end of the Section 2.3. The merger and ring-down parts of the signal are described by analytic functions with coefficients obtained from the fitting to the NR data. The advantages of these waveforms are obvious: they are very fast to generate and they can be extended/improved as more NR waveforms become available. One can see these models as an analytic fit to the available NR waveforms with the interpolation in between. The EOB waveforms are significantly slower to generate, as they require integration of Hamilton's dynamical equations, but they give us a physical insight into the problem of generation of GWs.

In Figure 2.5 (taken from [26]), we show a comparison between a publicly available NR waveform generated by the SXS collaboration and the EOBNR model for a precessing binary.

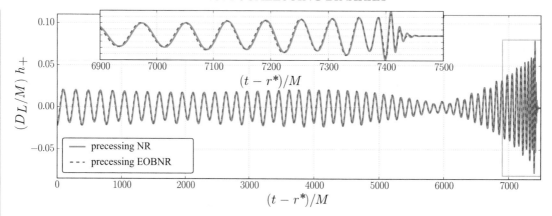

Figure 2.5: Comparison of NR h_+ and EOBNR for a binary BH with the mass ratio 5. The spin of the largest BH is $a = 0.5m_1^2$ and it initially lies in the orbital plane. The inset on the top is a zoom around the merger.

The blue is the NR waveform (h_+) and the dashed red line is the precessing EOBNR waveforms, parameters of the binary are given in the caption. One can see the modulation of the signal caused by the spin-orbital coupling: the orbital angular momentum precesses around the total momentum of the system. This waveform contains only $\ell = 2$, $m = \pm 2$, ± 1 modes.

The NR waveforms are computationally very expensive, and the longer the waveform the more time it takes to generate it, the same statement applies to systems with high spins and high mass ratio, therefore, the prime aim of NR waveforms is the calibration of analytical models described in this and previous sections. NR data could also be useful to understand the limitations of the analytical models especially in the ultra-relativistic regime close to the merger where we can potentially extract new physical phenomena.

The first observed GW event GW150914 is an exception; it has a relatively high mass and the parameter estimation shows that the BHs spins are rather small. This is the region of the parameter space where we have a relatively large number of NR waveforms and some simplified analysis of the data could be done with already existing NR templates. In general, for the search and parameter estimation, we need a set of waveforms with a good (fine) coverage of large parameter space and fast-to-generate GW templates (waveforms); this will be explained in the next chapter where we describe the data analysis algorithms.

CHAPTER 3

Gravitational Waves Data Analysis

We start with remembering the basic statistical notions, then we will give a brief overview of the frequentist and Bayesian approaches. We apply this theory to a problem of GW data analysis; there we overview common techniques used in different frequency bands—LIGO, LISA, PTA.

3.1 BASIC NOTIONS

In this section we want to remind the reader of some basic notions of the statistics and introduce notations which we will use in this chapter.

Probability is a way of assigning a number to an event which reflects the likelihood that this event will occur. If we have a complete set of possibilities then the total probability of this set is equal to unity, which means that at least one event out of the set happens, so the probability varies between zero and one. The events are called mutually exclusive if the appearance of one event excludes the possibility of another one. The important role play probability when two events A, B occur on a single measurement: $P(A \cap B)$. The two events are independent if $P(A \cap B) = P(A)P(B)$, whereas the probability of mutually exclusive events $P(A \cup B) = P(A) + P(B)$. An important role in the statistics plays Bayes' rule:

$$P(A \cap B) = P(B)P(A|B) = P(B \cap A) = P(A)P(B|A) \tag{3.1}$$

which could be rewritten as

$$P(A|B) = \frac{P(B|A)P(A)}{P(B)}. \tag{3.2}$$

Here we have introduced the conditional probability $P(A|B)$—the probability of the event A given that the event B has happened. Equation (3.2) could be interpreted as the probability of an event A has changed from the "prior" value $P(A)$ once we had additional information given by $P(B|A)$ which in this case is called a likelihood. The denominator $P(B)$ serves here as a normalization factor.

Another important set of notions is connected to the probability density function (pdf) which is defined as probability of a random variable x to happen in a infinitesimal interval dx:

$$P(x \in [x_0, x_0 + dx]) = p(x_0) dx. \tag{3.3}$$

We can define probability of a random variable x to fall in a finite interval $[a, b]$ as

$$P(x \in [a, b]) = P(x \leq b) - P(x \geq a) = \int_a^b p(x)dx.$$

This leads to the definition of the cumulative distribution function also often used in the statistical analysis:

$$C(x_0) = P(x \leq x_0) = \int_{-\infty}^{x_0} p(x)dx.$$

Finally, the normalization of the probability to unity implies

$$\int_{-\infty}^{+\infty} p(x)dx = 1.$$

In practice, we usually deal with the discrete sampling set and we approximate the pdf by a histogram, which is approximation of Eq. (3.3) with a final size bin $dx = \Delta x$.

The problem of detecting a signal in noise can be seen as a statistical hypothesis testing: given an observation/data we need to formulate a hypothesis (or a model describing the data) and see if the data supports it. For example, we have two hypotheses: $\mathcal{H}_0$ (null) hypothesis that the data is just Gaussian noise, and, $\mathcal{H}_1$ hypothesis that the data consists of Gaussian noise and, say, a deterministic signal. We will use expression "hypothesis testing" and "model fitting/selection" interchangeably as they refer here to the same thing. Almost always a model (hypothesis) is characterized by a set of parameters: $\{\lambda_i\}$. The models might or might not have a common subset of parameters.

There are two main approaches to hypothesis (model) testing.

Frequentist approach: here we treat a signal and parameters characterizing it as fully deterministic and the noise corrupts the signal. We try to determine if the signal is present in the data with a certain probability, and, if it is present, is characterized by a set of parameters $\{\hat{\theta}\}$ which is called "estimator." The "maximum likelihood estimator" is the most common, and, as the name says, this is the estimator which maximizes the likelihood function. The key in the frequentist hypothesis testing is the Newman–Pearson criterion; it maximizes the probability of the detection subject to a chosen significance of the test. In particular, it suggests that the likelihood ratio is the most powerful detection statistics in Gaussian noise. What it actually says is that we should start with fixing/choosing an acceptable false alarm probability (probability of taking $\mathcal{H}_1$ when $\mathcal{H}_0$ is true). Given a detection statistic y, the false alarm probability is computed as

$$P(y \geq y_{thr}|\mathcal{H}_0) = \alpha. \tag{3.4}$$

This defines a threshold y_{thr} and allows one to compute the detection probability $P(y \geq y_{thr}|\mathcal{H}_1)$. As mentioned above, the likelihood ratio is a powerful detection statistic which is defined as

$$\frac{P(d|\mathcal{H}_1)}{P(d|\mathcal{H}_0)} = \frac{P(d|\mathcal{H}_1, \theta_i)}{P(d|\mathcal{H}_0, \lambda_j)} \geq y_{thr}, \tag{3.5}$$

where d denotes the measurement data. It is usually the case that we do not know parameters of the signal $\vec{\theta}_i$ and sometimes we also need to model the noise, $\vec{\lambda}_j$, so the problem can be reduced to finding the threshold (assuming properties of the noise) and to the optimization (finding a maximum of the likelihood ratio in the multidimensional parameter space).

For a given noise realization, the maximum likelihood estimator of parameters $\vec{\theta}_{ML}$ might not be very close to the true values of a signal. How close the estimator is to the true values depends on a noise realization and the strength of the signal (signal-to-noise ratio)—the stronger signal is less "influenced" by a noise. The maximum likelihood is an unbiased estimator: the estimated values averaged over the noise realizations are approaching the true parameters.

Bayesian approach: In this approach we assign "the cost" to our decision, that is probabilities of occurrence of $\mathcal{H}_0$ and $\mathcal{H}_1$ which are called *priors*; π_0, π_1, before considering the outcome of measurements. We treat parameters characterizing the model as random variables, and, given the data and the model, we infer their probability density function, *posterior* pdf:

$$P\left(\mathcal{H}_i|d\right) = P\left(d|\mathcal{H}_i\right) \frac{\pi\left(\mathcal{H}_i\right)}{P(d)}. \tag{3.6}$$

Here, $P(\mathcal{H}_i|d)$ is a posterior probability, $P(d|\mathcal{H}_i)$ is the likelihood function, $\pi(\mathcal{H}_i)$ is a prior, and $P(d)$ is a normalization constants (posterior marginalized over all possible models). The likelihood describes measurement uncertainties; the prior reflects our knowledge/believes about $\mathcal{H}_i(\vec{\theta})$ for a set of parameters $\vec{\theta}$.

Consider several models $\mathcal{M}_i$ each parameterized by its own set of parameters $\vec{\theta}_i$ (the index i refers to a particular model). We want to assess which model is better supported by the observations (measurements), in other words, given the data d we need to evaluate the probability

$$P\left(\mathcal{M}_i|d\right) = \frac{P\left(d|\mathcal{M}_i\right) \pi\left(\mathcal{M}_i\right)}{P(d)} \tag{3.7}$$

and for a given model the posterior pdf is

$$p\left(\vec{\theta}_i|d, \mathcal{M}_i\right) = \frac{p\left(d|\vec{\theta}_i, \mathcal{M}_i\right) \pi\left(\vec{\theta}_i\right)}{p\left(d|\mathcal{M}_i\right)}. \tag{3.8}$$

The evidence of the model $\mathcal{M}_i$ is given by

$$p\left(d|\mathcal{M}_i\right) = \int d\vec{\theta}_i \, p\left(d|\vec{\theta}_i, \mathcal{M}_i\right) \pi\left(\vec{\theta}_i\right), \tag{3.9}$$

where the integral is evaluated over the whole parameter space of a model $\mathcal{M}_i$, and we can write

$$P\left(\mathcal{M}_i|d\right) = \left[\int d\vec{\theta}_i \, p(d|\vec{\theta}_i, \mathcal{M}_i)\pi(\vec{\theta}_i)\right] \frac{\pi(\mathcal{M}_i)}{P(d)}. \tag{3.10}$$

The biggest problem is evaluating the probability $P(d)$. It is possible to do so in the rare cases where the set of considered models is exhaustive and the models are mutually exclusive $P(d) = \sum_i P(d|\mathcal{M}_i)\pi(\mathcal{M}_i)$, but unfortunately those conditions are hardly ever met and we need to use a different method for selecting a model. Usually we consider the posterior odd ratio (relative probability), which allows us to avoid computing a normalization:

$$O_{ij}(d) = \frac{P(\mathcal{M}_i|d)}{P(\mathcal{M}_j|d)} = \left[\frac{\int d\vec{\theta}_i\, p\left(d|\vec{\theta}_i, \mathcal{M}_i\right)\pi\left(\vec{\theta}_i\right)}{\int d\vec{\theta}_j\, p\left(d|\vec{\theta}_j, \mathcal{M}_j\right)\pi\left(\vec{\theta}_j\right)}\right]\frac{\pi(\mathcal{M}_i)}{\pi(\mathcal{M}_j)}, \tag{3.11}$$

and, as you can see, we got rid of $P(d)$. The first term on the r.h.s. (fraction in the square brackets) is called the *Bayes factor* and the second fraction is called prior odds. If we do not have the prior preference of one model over another we assign the equal probability to each model. Note that the Bayes factor is somewhat related to (but not the same as) a likelihood ratio. If the priors are non-informative (flat), then the main contribution to each integral is defined by a likelihood function, so the integrals are dominated by the areas of parameter space with the largest likelihood. We have somewhat simplified the model selection, but still we need to compute those integrals, which are often multi-dimensional and have to be evaluated numerically. For further reading on the Bayesian approach, we refer to a very nice book [124].

3.2   DETERMINISTIC GW SIGNAL IN GAUSSIAN NOISE

Here we will consider a particular problem of detecting deterministic signal in Gaussian noise with zero mean. In particular, we will take a GW signal from coalescing compact binaries as a deterministic signal. We start with a simple example of Gaussian white noise and then extend it.

Consider Gaussian white noise with zero mean and variance σ. Usually we deal with the discrete measurements, $\vec{d} = \vec{n} = n(t_i)$, where, for simplicity, we assume the even sampling, $t_i = i\,\Delta t$. The assumption of the white noise implies that the samples are uncorrelated, $P(n(t_i)|n(t_{i\neq j}) = P(n(t_i))$ and the probability of observing N samples at different times is a product of probabilities at each instance:

$$P\left(\vec{n}\right) = \prod_{i=1}^{N} P\left(n\left(t_i\right)|\mathcal{H}_{wGn}\right) = \frac{1}{(\sigma\sqrt{2\pi})^N} e^{-\frac{1}{2\sigma^2}\sum_i n(t_i)^2}.$$

Here $\mathcal{H}_{wGn}$ explicitly shows that we assume white Gaussian noise. Note that the argument of the exponent looks like a weighted inner product of a vector $\vec{n}$ with itself. Introduce a general definition of the inner product which also works for a non-white noise:

$$<x|y> = \int_{-\infty}^{+\infty} df\, \frac{\tilde{x}(f)\tilde{y}^*(f) + \tilde{x}^*(f)\tilde{y}(f)}{S_2(f)} = 4\Re \int_{0}^{+\infty} df\, \frac{\tilde{x}(f)\tilde{y}^*(f)}{S(f)}, \tag{3.12}$$

where we have introduced one (two) sided power spectral density of the noise $S(f)$ ($S_2(f)$) and we operate in the frequency domain (here tilde denotes the Fourier transormed quantities):

$$\tilde{x}(f) = \int x(t)e^{-i2\pi ft} \, dt.$$

Indeed, in case of the white Gaussian noise the $S(f)$ is constant ($S_2(f) = \sigma^2$) and (3.12) corresponds to the zero-lag correlation in the time domain which is exactly what we have above. The power spectral density plays a role of the weight-function reflecting that the measuring device is not equally sensitive across the frequency band. Even if the signal has some non-trivial content at some frequencies we might not be able to detect it because of the high level of noise. Note that, if $x = d$ is data and y is a signal, then the inner product describes the *matched filter*. Returning back to the white Gaussian noise we can write the likelihood as

$$P\left(\vec{d} = \vec{n}|\mathcal{H}_0\right) \propto e^{-\frac{1}{2}<n|n>} = e^{-\frac{1}{2}<d|d>}. \tag{3.13}$$

This expression is valid beyond the assumption of the white noise if we use a general definition of the inner product and we have reinstated explicitly assumption of a null hypothesis, $\mathcal{H}_0$, that $\vec{d} = \vec{n}$.

Now consider the hypothesis that the data consists of Gaussian noise and a signal $\vec{d} = \vec{n} + \vec{s}(\vec{\theta})$. We can write the likelihood for this hypothesis, $\mathcal{H}_1$, as

$$P\left(\vec{d}|\mathcal{H}_1\right) \propto e^{-\frac{1}{2}<(d-s(\vec{\theta}))|(d-s(\vec{\theta}))>}, \tag{3.14}$$

which simply implies that, if we subtract the signal from the data, the residuals should be described by a Gaussian noise.

According to the Newman–Pearson lemma, the likelihood ratio is the optimal (the most "powerful") detection statistic from the frequentist point of view:

$$\mathcal{L}\left(d, \vec{\theta}\right) = \frac{P\left(\vec{d}|\mathcal{H}_1\right)}{P\left(\vec{d}|\mathcal{H}_0\right)} = \exp\left[< d|s\left(\vec{\theta}\right) > -\frac{1}{2} < s\left(\vec{\theta}\right)|s\left(\vec{\theta}\right) >\right]. \tag{3.15}$$

Introduce matched filter signal-to-noise ratio as

$$SNR^2 = < s|s >; \tag{3.16}$$

then taking the average over noise realizations of $\mathcal{L}$ and taking into account that the noise has zero mean, $\overline{d} = 0$, we have

$$\log \overline{\mathcal{L}} = \frac{1}{2} SNR^2. \tag{3.17}$$

Now we recall the result of the previous chapter where we have derived two polarizations of the GW signal, h_+ and $h_\times$. They are obtained in the frame associated with the source which we need to relate to the frame associated with the detector of GWs. We consider LIGO/Virgo-type detector, which is the L-shaped laser Michelson interferometer, and we place x, y axis along the arms, and the z axis is pointing to the zenith. The source sky location in this frame is given by two polar angles $\{\theta_S, \phi_S\}$ and the direction to the source is $-\vec{N} = -\{\cos \phi_S \sin \theta_s, \sin \phi_S \sin \theta_s, \cos \theta_s\}$, where $\vec{N}$ is a direction of GW propagation. In the source frame we have introduced the polarization basis $\partial_\theta, \partial_\phi$ given in the spherical polar coordinates (2.59), those are two vectors in the plane orthogonal to the vector $\vec{N}$. It is natural to introduce polarization basis in the detector frame in a similar way $\partial_{\theta_s}, \partial_{\phi_s}$, which is related to the basis in the source frame by a rotation angle ψ, called *polarization* angle. The response of the detector to the passing GW is given by a projection of GW strain h_{ij} on the arms of the detector, note that with the Michelson interferometer we measure the differential displacement, so that we have

$$h = \frac{1}{2} h^{ij} \left(x_i x_j - y_i y_j \right) \equiv F_+(t) h_+(t) + F_\times(t) h_\times(t), \tag{3.18}$$

where $F_{+,\times}$ are in general functions of time and they are given explicitly as

$$F_+ = \frac{1}{2} \cos 2\psi \, (1 + \cos \theta_s) \cos 2\phi_s - \sin 2\psi \cos \theta_s \phi 2\phi_s, \tag{3.19}$$

$$F_\times = \frac{1}{2} \sin 2\psi \, (1 + \cos \theta_s) \cos 2\phi_s + \cos 2\psi \cos \theta_s \phi 2\phi_s. \tag{3.20}$$

The functions $F_{+,\times}$ are called antenna response or antenna beam function. The time dependence comes through the motion of the detector (daily rotation, orbital motion), however, we can neglect this time dependence for short signals, which duration is much shorter than the characteristic time scale of a detector motion ($t_{sig} \ll T_d$). For a single detector the strain response can be written as

$$h(t) = \frac{A(t)}{D} \cos \left(2\phi(t) + \phi' \right), \tag{3.21}$$

where we have used Eq. (2.59), and the orbital phase $\phi(t)$ is represented by a PN Taylor series, we also assumed a short duration signal and

$$D = D_L \left[F_+^2 \left(1 + \cos^2 \iota \right) / 4 + F_\times^2 \cos^2 \iota \right]^{-1/2}, \tag{3.22}$$

$$\tan \phi' = \frac{2 F_\times \cos \iota}{F_+ \left(1 + \cos^2 \iota \right)}. \tag{3.23}$$

Here D_L is the luminosity distance, ι is inclination of the binary to the line of sight.[1] If motion of detector is negligible we cannot measure individually the distance, sky position and inclination, those parameters are combined together into what is called effective distance D and effective initial phase ϕ'. However, there are a few caveats: if we use GW model beyond restricted,

---

[1]See discussion on the inclination angle around Eq. (2.59).

namely, if we include sub-dominant harmonics, we can partially break this degeneracy. Different harmonics do not have the same dependence on the inclination and, if they are detectable, could lead to the estimation of the inclination. The same applies to the precessing binaries (if the spins are misaligned with the orbital angular momentum), then the inclination angle ι is a function of time and we potentially could measure both h_+ and $h_\times$. Finally, if the signal is long lived $t_{sig} \approx T_d$, the antenna response is time dependent (modulation of the amplitude and also Doppler modulation of the phase of a GW signal) which also breaks the degeneracy between parameters.

Having two or more detectors allows triangulation of the signal on the sky by measuring the relative time of arrival of a GW signal at different sites. Additional information is stored in the phase and amplitude, and the GW signal should appear with a coherent phase and amplitude at each detector.

3.2.1 $\mathcal{F}$-STATISTIC

As discussed above, within the frequentist approach we should use the likelihood ratio as a detection statistic, which we need to maximize over all parameters. For the GW signal the maximization over some parameters could be done analytically, and this is the basis for the $\mathcal{F}$-statistic. For simplicity, we start with the GW strain measured by a single detector. The strain could be written as

$$h = a_1 h_1 + a_2 h_2, \tag{3.24}$$

where $a_i = a_i(D, \phi')$ and

$$h_1 = A(t) \cos 2\phi(t), \quad h_2 = A(t) \sin 2\phi(t). \tag{3.25}$$

We will maximize the log of the likelihood ratio over the constants a_i, $i = 1, 2$:

$$\log(\mathcal{L}) = \sum_i a_i X_i + \sum_{i,j} a_i a_j M_{ij}, \tag{3.26}$$

where

$$X_i \equiv <d|h_i>, \quad M_{ij} =< h_i|h_j > . \tag{3.27}$$

We search for extremum (maximum) of the log-likelihood ratio with respect to the constants, $\partial(\log \mathcal{L})/\partial a_i = 0$, which gives us maximum likelihood estimator for the parameters:

$$a_j^{ML} = X_i M_{ij}^{-1}, \tag{3.28}$$

where we assume summation over the repeated indices. We can plug this estimator back to the log-likelihood to get

$$(\log \mathcal{L})_{\max a_i} = \frac{1}{2} X_i X_j M_{ij}^{-1}. \tag{3.29}$$

A particular case is if $M_{ij} = \delta_{ij}$; indeed, we can normalize the sub-templates so that $< \hat{h}_1|\hat{h}_1 >=< \hat{h}_2|\hat{h}_2 >= 1$, by moving the normalization constant into redefinition of $a_{1,2}$. Since cos and sin are orthogonal functions, we also have $< \hat{h}_1|\hat{h}_2 >\approx 0$, which is not entirely true because the signal has a finite duration but still gives us a good approximation, so we have

$$(\log \mathcal{L})_{\max_{a_i}} = \frac{1}{2}\left[< d \left|\hat{h}_1 >^2 + < d\left|\hat{h}_2 >^2\right.\right]\right., \tag{3.30}$$

this is sometimes referred to as computation of the log-likelihood in quadratures. We can generalize this to any signal which could be presented in the form

$$h = \sum_k^N a_k \hat{h}_k, \tag{3.31}$$

where N is a number of sub-templates $\hat{h}_i$. The maximization over a_k is analog to described above and we define the $\mathcal{F}$-statistic as

$$\mathcal{F} = \frac{1}{2} X_i X_j M_{ij}^{-1} \tag{3.32}$$

which was first introduced in $[85]^2$ and then further extended in $[25, 43]$. This is basically the log-likelihood ratio maximized over constants a_k. Let us find how it is related to the SNR:

$$SNR^2 =< h|h >= \sum_{ij} a_i a_j M_{ij}. \tag{3.33}$$

Consider the model $\mathcal{H}_1 : d = n + h$, and take an average over the noise realizations, then taking into account that $\bar{n} = 0$, $\overline{< n|\hat{h}_i >< n|\hat{h}_j >} = M_{ij}$,

$$\overline{X_i X_j} = M_{ij} + \sum_{kl} a_k a_l M_{ki} M_{lj} \tag{3.34}$$

and substituting this expression into (3.32) we get

$$\overline{\mathcal{F}} = \frac{1}{2}\left(N + SNR^2\right). \tag{3.35}$$

The $\mathcal{F}$-statistic is distribute as χ^2 with N degrees of freedom, whether it is central or not depends on presence of signal. We can use this statistical property to set the detection threshold for a desired false alarm probability. This statistic is actively used in the GW data analysis.

Another parameter which could be easily maximized over is a time of arrival of GW signal. This parameter is not uniquely defined, sometimes it is taken to be the instance of time when the GW signal reaches a certain frequency, alternatively, it is a time associated with the merger

---

^{2}Sometimes $2\mathcal{F}$ (in our notations) is called the $\mathcal{F}$-statistic.

(say, with the maximum of the amplitude of a GW signal), in the latter case it could be referred to as the time of coalescence. In any case, recall that the inner product is defined by (3.12), and we can extend it as

$$z(\tau) = \int_{f_{\min}}^{f_{\max}} \frac{\tilde{d}(f)\tilde{h}^*(f)}{S(f)} e^{-i2\pi f\tau} \, df, \tag{3.36}$$

which can be easily evaluated using a fast Fourier transform algorithm. This extension is equivalent to computing the inner product of a data with the time-shifted (by τ) template, and we can search for a maximum of the time series $z(\tau)$. For a discretely sampled data, the accuracy in determination of the coalescence time (using this method) is restricted by the sampling rate (besides the noise influence).

3.3   OPTIMIZATION METHODS

In the frequentist approach we want to find a maximum of the likelihood ratio. We have shown that for a certain form of GW template[3] we can analytically maximize the likelihood over some parameters. However, for other parameters we need to perform a numerical search for a maximum, this is often referred to as *optimization* problem. In this section we describe the most popular optimization methods used in the GW data analysis.

3.3.1   GRID-BASED SEARCH

Conceptually, the most straightforward method to find the global maximum of a function is to cover parameter volume by a uniform grid and evaluate the function (likelihood ratio in our case) at each point. The distance between the grid points should be chosen as a trade-off between the computational load and a danger to miss the maximum. There are two advantages of this method: (i) if the grid is fine enough, this method guarantees to find a global maximum of a likelihood and (ii) this method is trivially parallelizable and could be used hierarchically, starting with a relatively coarse grid and "zoom-in" to the most promising parts of the parameter space.

The main challenge of this method is to realize the uniform grid. The parameter manifold is not necessarily flat, and even if it is, it might be a non-trivial problem to find Cartesian-like coordinates. The distance between two nearby templates is defined via inner product

$$\begin{aligned} ds^2 &= \left\| \hat{h}\left(\vec{\theta} + \delta\vec{\theta}\right) - \hat{h}\left(\vec{\theta}\right) \right\| = \; < \hat{h}\left(\vec{\theta} + \delta\vec{\theta}\right) - \hat{h}\left(\vec{\theta}\right) \middle| \hat{h}\left(\vec{\theta} + \delta\vec{\theta}\right) - \hat{h}\left(\vec{\theta}\right) > \\ &\approx \left\| \frac{\partial \hat{h}}{\partial \theta_i} \delta\theta_i \right\| = \left\langle \frac{\partial \hat{h}}{\partial \theta_i} \middle| \frac{\partial \hat{h}}{\partial \theta_j} \right\rangle \delta\theta_j \delta\theta_i, \end{aligned} \tag{3.37}$$

---

[3]Template is a jargon used for GW model sampled in the parameter space. In matched filtering we search for a particular pattern of the signal buried in the noise using set of templates.

so the metric on the parameter space could be defined as

$$\gamma_{ij} = \left\langle \frac{\partial \hat{h}}{\partial \theta_i} \Big| \frac{\partial \hat{h}}{\partial \theta_j} \right\rangle. \tag{3.38}$$

Note that the grid has to be uniform in proper distance ds^2, and not in the coordinate distance (defined by a particular parametrization), which is hard problem even in 2-D. The distance ds tells us how similar (correlated) are two nearby templates. For a chosen distance $ds \ll 1$ we can define a volume of a template: this is the region in the parameter space with the (proper) distance from a given point less or equal to ds. Figure 3.1 shows the volume of a template O as a shaded ellipse in 2-D parameter space. The semi-major axis of this ellipse corresponds to the direction of a strong correlation (we need to move further away to have quite distinct templates) and the semi-minor axis corresponds to the direction of weak correlation (a small change in parameters in that direction leads to sizable changes in the template). Note that the size and orientation of the ellipse is the function of the central point O.

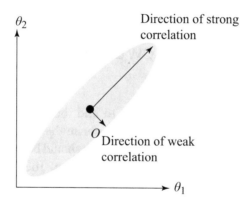

Figure 3.1: Volume of a template.

The grid is used in LIGO-Virgo GW data analysis, and it is constructed either by finding the combination of parameters along which the metric is smooth and slowly changing function, or using the stochastic method. In the stochastic method [21, 78], one chooses templates randomly (uniformly or according to some preferred distribution) within preset boundaries checking that every new point is not too close to ones already chosen. The collection of points in the parameter space (constructed grid) is called a *bank of templates*. The stochastic method has the problem of completeness (never can be sure that there are no "holes") and very low efficiency as we approach the final size of the bank.

Another idea suggested in [99] and further developed in [97] is a random placement: first one needs to estimate the number of required templates (by evaluating the total (proper) volume of the parameter space and dividing it by a characteristic volume of a single template). Once we know the number of templates we just randomly place them over all parameter volume.

Another problem of the grid-based search is that this method becomes prohibitively computationally expensive as we increase the number of dimensions. The signal from the black hole binary with spins not aligned with the orbital angular momentum is characterized by 15 parameters. It is hard (if possible at all) to cover uniformly 15-D manifold, in addition the number of templates grows roughly as a power-low with the number of dimensions.

As we conclude this section we want to address one more issue: *effectualness* and *faithfulness* of the template model. So far we have assumed that the GW model perfectly resembles the expected GW signal $s = h(\vec{\theta}_{true})$. In reality we use GW models which are only approximate solutions to the two-body problem (see the previous chapter). If $s \neq h(\vec{\theta}_{true})$, then we will have a systematic bias in the parameter estimation:

$$\min_{\vec{\theta}} \left\langle s - h\left(\vec{\theta}\right) \middle| s - h\left(\vec{\theta}\right) \right\rangle \rightarrow \tilde{\vec{\theta}}.$$

Note that the bias does not depend on the strength of the signal, while the statistical error (due to presence of noise) scales as inverse of the SNR. This is an important issue for the loud signals; the systematic error might start to dominate over a statistical error, which can be resolved only by improving the accuracy of a GW signal model.

Despite the bias in the parameter estimation $\delta\vec{\theta} = \tilde{\vec{\theta}} - \vec{\theta}_{true}$, we still might be able to detect the GW signal; in this case a GW model (template) is called *effectual*. The measure of effectualness is the *overlap*, $\mathcal{O}$, maximized over all parameters:

$$FF = \max_{\vec{\theta}}(\mathcal{O}), \quad \mathcal{O} = < \hat{s} | \hat{h}\left(\vec{\theta}\right) >,$$

which is also called *fitting factor*, *FF*. Note that the overlap is computed between two normalized models $\hat{s}$ and $\hat{h}$, ($< \hat{s}|\hat{s} >=< \hat{h}|\hat{h} >= 1$) so it varies in $[-1, 1]$, unity being a perfect match. In reality, we use NR waveforms as a proxy for the expected GW signal to assess "goodness" of a given model (approximant). The fitting factor could be high on the expense of the bias in parameters for not sufficiently accurate (not faithful) models. The "faithfulness" is another relative; it is an overlap between a signal (read, NR model) and the model template taken for the fix intrinsic parameters of a system. The intrinsic parameters are masses and spins of the binary system. The overlap could be maximized over extrinsic parameters (defining the relative orientation of the source frame and the detector). Faithfulness serves as an indirect proxy for the possible bias in the parameter estimation.

3.3.2   GENETIC ALGORITHM

Genetic algorithm is a widely used optimization technique. The main idea here is to evolve the population of points in the parameter space to the state with a better "quality." It is similar to Darwin's natural selection theory: the strongest organisms survive and most likely to participate in the reproduction process so that the "good" properties propagate to the next generation. Let us provide a dictionary which defines a correspondence between biological terms and data analysis notions.

- Organism: a point in the parameter space.

- Colony of organisms: a set of points/templates in the parameter space.

- Genes: parameters characterizing a template (organism), $\vec{\theta}_i$, where $i = 1 \ldots N$ numerates organisms in a colony.

- Fitness (or goodness): this is the logarithm of a likelihood ratio, $\log \mathcal{L}$, better organisms should have higher $\log \mathcal{L}$.

- Generation: current position (distribution) of points $\{\vec{\theta}_i\}^{(k)}$, we evolve $\{\vec{\theta}_i\}$ from kth generation to $k + 1$.

Evolution of a colony is defined by a set of rules for (i) selection for breeding, (ii) breeding, (iii) mutation, (iv) elitism, and (v) accelerators.

Selection: This is the rule to select two or more organisms (parents) for breeding. For example, one could use the "roulette rule," where we select organisms with a probability

$$P_i = \frac{\log \mathcal{L}(\vec{\theta}_i)}{\sum_j \log \mathcal{L}(\vec{\theta}_j)}.$$

Breeding: This is the rule to create one or more children from the selected parents. Let us show two examples of breeding from two chosen organisms characterized by parameters $\vec{\theta}_A$, $\vec{\theta}_B$. We can represent each parameter in a binary format and divide the digits in half (single point cross-over breeding rule). To form a "child" (organism "C") we take the first half from the parent "A" and the second half from the parent "B". It is not necessary to go to the binary format (the example of this method is illustrated in Figure 3.2). We also show two-cross-points breeding using the decimal representation in Figure 3.3. You can set your own rules and check their efficiency.

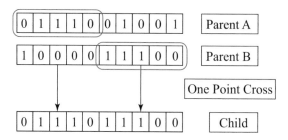

Figure 3.2: Breeding example 1: binary representation of parameter, one-point crossover.

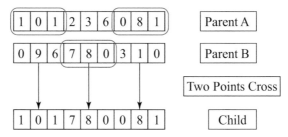

Figure 3.3: Breeding example 2: decimal representation of parameter, two-point crossover.

Mutation: This is the rule which allows changing a gene (parameter) of an organism with some probability, and this probability is called the probability of mutation rate (PMR). Consider one example: we choose uniformly distributed random variable $\alpha \in U(0, 1)$ and if $\alpha < PMR$ we mutate a given parameter, for example, by flipping four randomly chosen bits in the binary representation. This example is given in Figure 3.4. Again, one can make up their own rules for mutation. Mutation, being a random process, could be "negative" or "positive," which leads to a decrease or increase in the log-likelihood of a mutated template. Like in nature the mutation could lead to the death of an organism or make it stronger (like the spiderman). The high value of PMR causes the frequent mutation and, as a result, allows to explore a large volume of parameter space. This is a desired feature when we search for a maximum in multi-dimensional space, while low PMR is more suitable for exploring a local area. The latest is suitable when we converge to a certain (maybe local) maximum. This suggests to make PMR a function of evolution, function of generation number, starting with a high PMR initially and decrease it as we evolve a colony.

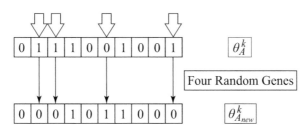

Figure 3.4: Example of mutation: binary representation of parameter, mutation of four random genes.

Elitism: (or cloning) Here we propagate the best current organism into the next generation. This guarantees convergence of a Genetic Algorithm to a solution; however, it does not guarantee that the solution is a global maximum. This is a general statement: a Genetic Algorithm being a stochastic search does not guarantee that the global maximum will be found; rather, the success depends on the implementation and on the number of generations before we terminate the evolution.

Accelerators: Those problem-tailored rules which exploit the properties of the signal and allow faster convergence. This is especially important for the multi-modal likelihoods (like an egg-box). In this case it is useful to evolve several colonies with different rules. Some colonies are exploring a large volume of parameter space and searching for maxima, while other colonies explore the space around each found maximum. We have found the Genetic Algorithm to be a very powerful tool with unlimited tuning options [105], which is also a problem as it is easy to get lost among those options.

3.3.3   PARTICLE SWARM OPTIMIZATION

This is another method borrowed from nature. I like describing it using a swarm of bees as an example. Every morning bees fly from a hive in search of a flower field. Each bee during the day performs random flights coming across "good" and/or "bad" spots in the field (many flowers or no flowers). So, the ith bee has found during the day its own "best spot;" in terms of data analysis, it has found the point in the parameter space $\vec{\theta}_i^b$ with the highest log-likelihood. In the evening all bees share information about their findings and they identify the "best global spot" $\vec{\theta}^g = \max_i(\log \mathcal{L}(\vec{\theta}_i^b))$. This information is updated every day, and the swarm uses the information accumulated by the day "k" during its path for the day $k + 1$. The speed (which defines the direction of motion) of each bee for the day $k + 1$ is given as

$$\vec{v}_{(i)}[k + 1] = w_k \vec{v}_{(i)}[k] + c_1 \chi_1 \left(\vec{\theta}_i^b[k] - \vec{\theta}_i[k]\right) + c_2 \chi_2 \left(\vec{\theta}^g[k] - \vec{\theta}_i[k]\right), \tag{3.39}$$

where w_k is a weight of velocity the bee had a day before, and in general it is an evolving function of k; c_1, c_2 are accelerations of motion toward its own and the global best spots found so far. The values $\chi_{1,2}$ are chosen randomly from the uniform distribution χ_1, $\chi_2 \in U[0, 1]$. The jump from the current position of a template (bee) is updated with new values of velocity. We do not want a bee to fly to the moon, so we should restrict the value of velocity to some maximum value $|\vec{v}_i| < v_{max}$. Usually, the search parameter space has boundaries, either physical or imposed by some other considerations. There could be three types of boundary conditions for the swarm's evolution.

- We set the likelihood to zero outside the restricted parameter space, and let the particle (bee) to fly back to the regions with a high likelihood.

- We bounce the particle back (reflection) by changing the sign of the velocity component orthogonal to the boundary $\vec{v}_\perp \rightarrow -\vec{v}_\perp$.

- We can impose absorbing boundary by setting the $\vec{v}_\perp = 0$ at the boundary and beyond.

The big advantage of this method is that it is very easy to implement and use. The biggest problem is convergence and stopping conditions. The evolution should be stopped when all particles (bees) are clustered in a small volume of parameter space. As for finding a global maximum, it helps to run the search several times with different initial state. This approach was used in GW data analysis several times [131, 148] and proved to be relatively successful for rather simple likelihood surfaces.

3.4   BAYESIAN APPROACH

We want to compute the Bayes factor to assess which model is preferred; that means we need to evaluate a multi-dimensional integral

$$p(d|\mathcal{M}) = \int p\left(d|\vec{\theta},\mathcal{M}\right) \pi\left(\vec{\theta}|\mathcal{M}\right) d\vec{\theta}. \tag{3.40}$$

Only in a very rare occasion can this integral be evaluated analytically in full; sometimes it is possible to marginalize (integrate) it over some parameters, but in general, we need to perform integration numerically. There are several techniques to do so which we will describe below, and the main idea is to find the regions of the parameter space where the integrand (in absolute value) is the largest, so the contribution to the integral from those parts is dominant. Usually this corresponds to the regions with a large likelihood (however the prior, if not flat, could give also a significant contribution). We could use the techniques of the previous section to find the parts of parameter space where the integrand is large and vice versa; the methods described in the following sections could be used to find the maximum of the likelihood. We will consider three methods for evaluating integral numerically: vegas [93], Markov chain Monte-Carlo (MCMC) [70], and the nested sampling [125].

3.4.1   VEGAS ALGORITHM

This method is based on the importance sampling (identification of the regions where the integrand is large). The approximation to the integral could be written as

$$I(f) = \int_V f\left(\vec{x}\right) d\vec{x} \approx \frac{V}{N} \sum_i f\left(\vec{x}_i\right) \equiv V\overline{f}, \tag{3.41}$$

where V is the volume of the parameter space, N is a number of samples $\{\vec{x}_i\}$, $i = 1 \dots N$ at which we evaluate the integrand, and $\overline{f}$ is the mean value of integrand over the chosen samples. The error in the integral is given as

$$\sigma_I = V \sqrt{\frac{1}{N}\left(\overline{f^2} - \overline{f}^2\right)}, \tag{3.42}$$

Ideally, we want to draw more samples in the part of parameter space where the contribution to the total integral is largest by choosing an appropraite pdf $g(\vec{x})$. The probability distribution function $g(\vec{x})$ should be also limited to the integration volume: $\int_V g(\vec{x}) d\vec{x} = 1$. We can write the integral as

$$I(f) \; = \int_V f(\vec{x}) \, d\vec{x} = \int_V \frac{f(\vec{x})}{g(\vec{x})} \, g(\vec{x}) \, d\vec{x}$$

$$\approx \overline{\frac{f}{g}} \pm \sqrt{\frac{1}{N} \left[\overline{\left(\frac{f}{g}\right)^2} - \left(\overline{\frac{f}{g}}\right)^2 \right]} \tag{3.43}$$

and search for $g(\vec{x})$ which minimizes the variance (under the square root):

$$g = \frac{|f|}{I(|f|)}, \tag{3.44}$$

where the denominator is the integral of $|f(\vec{x})|$. This expression implies that g should resemble closely the integrand $|f|$. This pdf could be found iteratively, we start with the uniform sampling: we divide the parameter space into equal hypercubes (boxes) and draw randomly equal number of points in each box. Then we compute g_0 according to (3.44). In the next step, we re-adjust the size of each box according to the density g_0, so that the boxes are smaller in the regions of high values of the integrand, and larger where the contribution to the integral is small. Then we repeat the procedure drawing equal number of points in each box and form g_1. The iterations are stopped when the required precision of the integral is reached; more details can be found in [93], and this method is implemented in the publicly available "gnu scientific library." One can also see this method as a mesh refinement, or adjusting the bins of the (multidimensional) histogram (but preserving the normalization) to get a required pdf. The advantage of this method is that it gives us control on the error of the integral and one can reach the required precision. However, the method becomes very (computationally) expensive as we increase the number of dimensions: we require more boxes and, probably, more points per box so that we do not miss sharp features in the integrand (in the likelihood). This method is well parallelizable since we can compute the likelihood in each box independently.

3.4.2   MARKOV CHAIN MONTE CARLO (MCMC)

This method is a Monte Carlo integration with the help of a Markov chain. The chain is constructed in such a way that it spends most of the time in the part of the parameter space with a high likelihood. Markov chain is a stochastic process where the next point in the chain $\vec{\theta}_{k+1}$ depends only on the previous one $\vec{\theta}_k$. Formally, we call "Markov chain with a transitional probability $P(\vec{\theta}_{k+1}|\vec{\theta}_k)$" if there is a transitional probability $P(\vec{\theta}_{k+1}|\vec{\theta}_k)$ which brings the chain from the state $\{k\}(\vec{\theta}_k)$ to the state $\{k+1\}(\vec{\theta}_{k+1})$ and it depends only on kth state. If the Markov chain satisfies the balance equation:

$$\Lambda\left(\vec{\theta}_k\right) P\left(\vec{\theta}_{k+1}|\vec{\theta}_k\right) = \Lambda\left(\vec{\theta}_{k+1}\right) P\left(\vec{\theta}_k|\vec{\theta}_{k+1}\right), \tag{3.45}$$

then the chain starts to sample a distribution $\Lambda(\vec{\theta}_k)$ after some "*burn-in*" length. In our case we want to sample a posterior distribution function. Markov Chain Monte Carlo (MCMC) is widely used in the problems where we want to estimate parameters of the signal. Here we consider a particular realization of MCMC, namely the Metropolis–Hastings algorithm [79]. The Metropolis–Hastings realization suggests a particular way of building the transitional probability which satisfies the balance equation for the posterior pdf. First, we need to introduce a proposal distribution $q(\vec{\theta}_{k+1}|\vec{\theta}_k)$, which, in principle, could be arbitrary, but the algorithm is most efficient if this proposal distribution resembles closely the expected posterior pdf. Introduce the acceptance probability α such that

$$\alpha\left(\vec{\theta}_{k+1},\vec{\theta}_k\right) = \min\left\{1, \frac{p\left(d|\vec{\theta}_{k+1}\right)q\left(\vec{\theta}_k|\vec{\theta}_{k+1}\right)\pi\left(\vec{\theta}_{k+1}\right)}{p\left(d|\vec{\theta}_k\right)q\left(\vec{\theta}_{k+1}|\vec{\theta}_k\right)\pi\left(\vec{\theta}_k\right)}\right\}, \tag{3.46}$$

where the last term is called Metropolis-Hastings ratio which insures the balance equation (3.45). We also have dropped $\mathcal{M}$ in $p(d|\vec{\theta}_k) \equiv p(d|\vec{\theta}_k,\mathcal{M})$ which is the likelihood function and $\pi(\vec{\theta}_k)$ is a prior pdf. The transitional probability of the Markov chain is $P(\vec{\theta}_{k+1}|\vec{\theta}_k) = \alpha(\vec{\theta}_{k+1},\vec{\theta}_k)q(\vec{\theta}_{k+1}|\vec{\theta}_k)$. In practice, we start with a first point $\vec{\theta}_0$ in the chain chosen randomly according to the prior pdf and compute the likelihood at that point. We choose the next point in the parameter space according to the proposal $q(\vec{\theta}_1|\vec{\theta}_0)$ and compute the prior, $\pi(\vec{\theta}_1)$, and the likelihood, $p(d|\vec{\theta}_1)$, in a new point. Then we evaluate the Metropolis–Hastings ratio, and, if it is larger than 1, accept the new point. If the ratio is less than one, we draw $\beta \in U[0,1]$ and, if $\beta > \alpha$, accept the new point with the probability β and reject it otherwise keeping (adding again) the present point and propose the next jump. Even though the MCMC is a stochastic process it moves predominantly toward the spots in the parameter space with a large likelihood, and after some time (called burn-in stage) it settles on around regions with high values of the integrand (of evidence) and samples the posterior pdf.

The MCMC can easily get stack at the local (secondary) maxima, while there is non-zero probability that it will move off and find other maxima, in reality it might take a really long (unacceptably long) time. There are several additional techniques which ensure the efficient exploration of the parameter space, here we consider the most widely used one called *simulated annealing*. Let us rewrite the likelihood ratio as

$$\mathcal{L} \propto e^{<d|h(\vec{\theta})-\frac{1}{2}<h|h>} \equiv e^{\frac{<d|h(\vec{\theta})-\frac{1}{2}<h|h>}{T}} \equiv e^{-\frac{E(\vec{\theta})}{T}}, \tag{3.47}$$

where $T = 1$. We notice that the last form of the likelihood ratio reminds the Boltzmann distribution (distribution of particles in a system over the energy states $\propto e^{-E_i/kT}$), where T is a temperature. The idea of the simulated annealing is to "heat up" the likelihood surface by using the high temperature $T > 1$. The high temperature makes the likelihood smoother by increasing the noise level, so we absorb the small maxima and maxima become wider, this increases the acceptance rate and the chain efficiently explores a large parameter space which helps to find a global maximum. However at the end we want the results from the chain with $T = 1$, so we can start a run with a high temperature and cool it down to unity as we progress, $T = T(k)$ or $T = T(\mathcal{L}(\vec{\theta}_k))$. The simulated annealing has proven to be quite powerful.

The MCMC gives us posterior pdf (which is very useful) and it is used to assess parameters of detected coalescing binaries, but not the evidence directly and that is what we also want. We can use a technique called *parallel tempering* where we run multiple chains in parallel with different temperatures and make them "talk" to each other. This method allows us to explore the parameter space at different scales: large scale exploration with high temperature chains, exploration of parameter space around maxima with the low-temperature chains. We perform cross-talk between the chains by exchange of the points in the chains $\vec{\theta}_k^{T_i} \leftrightarrow \vec{\theta}_k^{T_j}$ (or, equivalently, swap the temperature) with probability

$$p = \min \left\{ 1, e^{(E_i - E_j)\left(\frac{1}{T_i} - \frac{1}{T_j}\right)} \right\}, \tag{3.48}$$

where the energy is introduced in Eq. (3.47). In addition, this method helps us to evaluate the evidence integral: introduce a new variable $\beta = 1/T$, the evidence corresponding to each chain with the temperature T is

$$Z(\beta) = \int d\vec{\theta} \mathcal{L}\left(\vec{\theta}, T\right) \pi\left(\vec{\theta}\right). \tag{3.49}$$

We are interested in the chain with $T = 1$, that is $Z(1)$. First, we differentiate the integral above with respect to β

$$\frac{\partial \log Z(\beta)}{\partial \beta} = \frac{1}{Z(\beta)} \int d\vec{\theta} (\log \mathcal{L}) \mathcal{L}\left(\vec{\theta}, \beta\right) \pi\left(\vec{\theta}\right) = \left\langle \log \mathcal{L}\left(\vec{\theta}\right) \right\rangle_{\beta}, \tag{3.50}$$

where the last equality means the average of the $\log \mathcal{L}(\vec{\theta})$ over the posterior at $T = 1/\beta$. Now we can express the evidence by integrating the above expression (note that $Z(0) = 1$):

$$\log Z(1) = \int_0^1 \left\langle \log \mathcal{L}\left(\vec{\theta}\right) \right\rangle_{\beta} d\beta. \tag{3.51}$$

In practice, we evaluate the average $\log \mathcal{L}(\vec{\theta})$ for each chain and apply the quadrature formula to evaluate the integral. It is important to cover well the range of temperatures (β) to maintain the error in the integral low, and, usually, a uniform distribution of $\log T$ is a good choice.

The important part of MCMC is a *convergence*: when should we stop the chain? First we should note that not all the points in the constructed chain are independent, and it is the number of independent points define the actual length of the chain. However, the independence of points might not be important for plotting histograms of the posterior. We use autocorrelation to evaluate the characteristic length over which we can consider the points to be uncorrelated and it can be used to down-sample the chain. The autocorrelation of the data $\{x_i\}$ is defined as

$$R(l) = \sum x_n x_{n-l}^*, \tag{3.52}$$

where we assume that the data could be complex, or in continuous process we have $R(\tau) = \int x(t)x^*(t - \tau)d\,t$. There are several definitions of the characteristic autocorrelation length: (i) l_{char} such that the autocorrelation drops by e: $R(l_{char}) = 1/e$ assuming that the autocreelation at zero lag is normalized to 1; and (ii) l_{char} such that the autocorrelation drops to 0.2. The autocorrelation length l_{char} tells us that roughly every l_{char}th sample is independent and it gives us the actual length of a chain. The autocorrelation length could be decreased by improving the jump proposal distribution $q(\vec{\theta}_{k+1}|\vec{\theta}_k)$. Note that the auto-correlation length could be different for each parameter.

The chain could stack at local maximum, using simulated annealing and parallel tempering helps a lot to explore the parameters space. It is also a good practice to run several chains in parallel and check how similar they are. For several chains we could use Gelman–Rubin [70] criterion to check convergence. This criterion is based on the variance of each chain (for each parameter) and on the dispersion between the chains (for that all chains should use temperature $T = 1$). It also helps to look at the chains. Another possibility is to use predictive distribution, which is trying to answer the question: if I have a model and posterior samples, is it any good? Assume we have more data d', the posterior predictive distribution is $p(d'|d)$:

$$p\left(d'|d\right) = \int d\,\vec{\theta}\, p\left(d'|\vec{\theta}\right) p\left(\vec{\theta}|d\right),\tag{3.53}$$

where we use the previously obtained posterior pdf $p(\vec{\theta}|d)$ as a prior to check the new data.

There is another way to evaluate the Bayes factor which is applicable if the models are nested: the model $\mathcal{M}_0$ is a particular case of a model $\mathcal{M}_1$ when $\vec{\theta} = \vec{\theta}_0$, in this case the Bayes factor could be estimated as

$$B = \frac{p\left(\vec{\theta} = \vec{\theta}_0|d\right)}{\pi\left(\vec{\theta} = \vec{\theta}_0\right)}.\tag{3.54}$$

It tells us how much the posterior deviates from the prior at the point where the model $\mathcal{M}_1$ reduces to the model $\mathcal{M}_0$. We will stop here description of MCMC and refer reader to other (statistical) literature specifically dedicated to this subject (for example, see [70]).

3.4.3   NESTED SAMPLING

We will give only a very brief overview of the nested sampling, technique suggested by Skilling [125]. We convert the integral for the evidence

$$p(d|\mathcal{M}) = \int p\left(d|\vec{\theta}, \mathcal{M}\right) \pi\left(\vec{\theta}|\mathcal{M}\right)\,d\vec{\theta}.$$

to the 1-D integral, to do that we introduce the prior (probability) mass function $dX = \pi(\vec{\theta})\,d\vec{\theta}$ and

$$X(\lambda) = \int_{L(\vec{\theta})>\lambda} dX = \int_{L(\vec{\theta})>\lambda} \pi\left(\vec{\theta}\right)\,d\vec{\theta}.\tag{3.55}$$

In this section we drop $\mathcal{M}$ assuming that we are considering a particular model, and we use the notation $L(\vec{\theta}) = p(d|\vec{\theta}, \mathcal{M})$ for the likelihood function.

Note that $X \in [0, 1]$ because the priors describe probabilities, and $L(X) = L(X(\lambda)) \equiv \lambda$ is an inverse function; in addition, $L(X) > 0$ and monotonically decreasing. The evidence could be written as

$$Z = \int_0^1 L(X)\,dX. \tag{3.56}$$

The main idea is to find the points $0 < X_M < \ldots < X_2 < X_1 < X_0 = 1$ which correspond to $L_0 < L_1 < \ldots < L_M$, so we can evaluate the integral above using, say, trapezoid sum. This integration procedure is very close to the Lebesque technique where we slice the integrand not in vertical (like in the Riemann integration) but in the horizontal stripes.

In practice, we start with N live points in the parameter space chosen arbitrarily from the prior and initiate $Z = 0$, $X_0 = 1$. Those live points will serve to explore the likelihood surface. At each iteration "j" we identify the live point with the lowest likelihood $(L_j = \min_{\vec{\theta}_i} (L(\vec{\theta}_i))) \rightarrow \vec{\theta}_j)$. We associate this point with $X_j = e^{-j/N}$ and we replace the point $(\vec{\theta}_j)$ with the lowest likelihood with a new point such that $L(\vec{\theta}^{new}) > L_j$ in proportion to the prior $\pi(\vec{\theta})$. The last step requires to search parameter space to find a region with a high likelihood, and one can use any method described above (genetic algorithm, particle swarm optimization, MCMC) to do so. We can evaluate the evidence integral at each iteration and terminate the iterations when the integral does not increase anymore (within some tolerance). Out of the selected points $\vec{\theta}_j$ we can construct the posterior by resampling the points with weights $p_j = (L_j w_j)/Z$, $(w_j = X_j - X_{j+1})$. This method has proven to give quite a robust evaluation of the evidence.

3.5   NON-GAUSSIAN NOISE

The problem with non-Gaussian noise is that there are infinite possibilities for the noise to be non-Gaussian. In LIGO/Virgo data analysis we search for the source of non-Gaussianity, those are various non-stationary instrumental artifacts, and try to remove/fix it. Unfortunately, not all non-stationary noise can be easily identified and removed, there are still present various transient instrumental and environmental artefacts and we need to deal with them. While the previous (sub)sections were quite generic, here we concentrate on GW data analysis.

It helps a lot to have two or more detectors (which is the case of LIGO-Virgo, and the LISA observatory with three arms, which we will discuss in the next chapter). In this case we can demand the coincidence of the GW signal within the light travel time between the sites. Besides, the GW signal should appear coherent at different detectors: with the consistent phase, amplitude, and intrinsic (to the source) parameters. The coincidence/coherence test is by far the most powerful way to eliminate the artifacts, but of course, it requires two or more detectors with a similar sensitivity operating simultaneously.

In this section we will discuss the approach which allows differentiating between a GW signal and noise, it is often referred to as a signal based veto. It uses information about the signal to check if the candidate event possesses the same properties. The most obvious property of GW is its chirp-like structure. If we plot a GW signal from coalescing binaries on the time-frequency plane we will see the characteristic track described by Eq. (2.64) with the amplitude growing as $f^{2/3}(t)$. We could check if the candidate event has this (chirping) property. The glitches (transient noise artifacts) usually do not look like GW signals, however they could be very strong and even poor match could produce a significant SNR, indeed $SNR \propto < d|\hat{h}>$, where $\hat{h}$ is a normalized template and $d = n + g$ is a data which we have presented here as a Gaussian part plus a glitch, and $g = A_g \hat{g}$. A template and a glitch could have a low match $< \hat{g}|\hat{h}> \ll 1$, but it could be compensated by a large amplitude A_g. In this section we first introduce the χ^2 test suggested in [13] and then we generalize it and discuss other possibilities to deal with the non-stationary noise.

3.5.1   SIGNAL-BASED VETO: χ^2 TEST

This section is a summary of [13], and it is added here mainly to introduce the notations and basis for the generalization described in the next section.

Introduce a GW signal from a coalescing binary as

$$s = A \left[\cos (\phi_0) h_c (t - t_0) + \sin (\phi_0) h_s (t - t_0) \right] \tag{3.57}$$

and the template in frequency domain

$$\hat{h} = (h_c + i h_s) e^{-2i\pi f t_0} \quad < \hat{h}|\hat{h} >= 1. \tag{3.58}$$

Here we consider a single detector, ϕ_0 plays role of an initial phase as seen in the detector frame, and t_0 is some characteristic arrival time of a GW signal. We denote the data as $d = n + s$, where n is Gaussian noise with zero mean.[4] Introduce

$$z \equiv < \hat{h}|d >, \quad \bar{z} = \overline{< \hat{h}|d >} = A e^{i\phi_0}, \tag{3.59}$$

where the over-bar means as before an average over the noise realizations, and we have assumed that the template and a signal are "identical" (no systematic bias, no mismatch in the parameters). Note also that $\overline{z^2} = 2 + A^2$. The quantity z serves as a proxy for the SNR (see Section 3.2.1). Introduce the incremental SNR (z_i) over the frequency band $[f_i, f_i + \delta f_i]$

$$z_i = 4\Re \int_{f_i}^{f_i + \delta f_i} \frac{\tilde{d}(f)\hat{h}^*(f)}{S_n(f)} df. \tag{3.60}$$

We choose the frequency bands δf_i such that each such bin gives approximately equal contribution to the total SNR: $< \hat{h}_c|\hat{h}_c >_i = < \hat{h}_s|\hat{h}_s >_i = q_i$ and $\sum_i q_i = 1$. This scheme is illustrated

---

[4]Here we want to differentiate the data $d = n + s$ from the data which contains a glitch $d = n + g$, assuming that in both cases the underlying noise is Gaussian.

in Figure 3.5 which shows how the cumulative SNR is split into the frequency bins of equal contribution. Note that the bins are usually not equal in size.

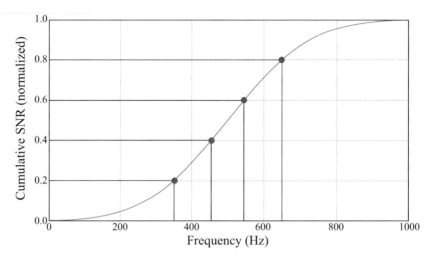

Figure 3.5: Schematic plot showing splitting the signal into the frequency bins which correspond to subtemplates with an equal contribution to the total SNR.

Now we compute the match between the data and template in those frequency bins and compare with the expected values:

$$\Delta z_i = z_i - q_i z, \tag{3.61}$$

where z_i is a measured match and $q_i z$ is an expected contribution. Introduce the quantity

$$\chi^2 = \sum_{i=1}^{p} \frac{\Delta z_i^2}{q_i}, \tag{3.62}$$

this quantity has χ^2-distribution with $2p - 2$ degrees of freedom (p is the number of frequency bins in our splitting). Note that even though we have chosen all q_i to be equal, it is not necessary. In reality, we could expect some mismatch between the signal and templates (systematic bias) and/or slight mismatch in the parameters (say due to coarseness of a template bank), which gives a correction to the expectation value

$$\overline{\chi^2} = 2p - 2 + \kappa |\bar{z}|^2 \tag{3.63}$$

which is proportional to the square of SNR. This implies that we might mistake high SNR events for a glitch if the systematic bias is not taken care of. The basis of this test is that we check that each sub-template (according to our splitting of a total frequency band into bins) should produce an equal (expected) SNR contribution in the matched filter. In the case of glitches, most SNR is usually concentrated in 1-2 particular bins and they fail this signal consistency test. This test has proven to be very useful for broadband signals.

3.5.2   SIGNAL BASED VETO: GENERIC TIME-FREQUENCY TEST

Here we will generalize the χ^2 test introduced above. We will be working in the time-frequency domain and introduce the following notations:

$$\Delta z\,(t_0, \tau, F) = z\,(t_0 \pm \tau, F) - P\,(\tau, F)\,z\,(t_0), \tag{3.64}$$

where t_0 is a time of arrival associated with a candidate event (say time of coalescence, t_c), $z\,(t_0) = z\,(t_0, F = f_{max})$ and

$$P(\tau, F) = 4\Re \int_{f_{min}}^{F} \frac{\hat{h}(f)\hat{h}^*(f)}{S_n(f)} e^{2\pi i f \tau}\, df. \tag{3.65}$$

If $\tau = 0$, $P(\tau = 0, F)$ gives us a cumulative SNR as a function of upper frequency cut-off. If $F = f_{max}$, where f_{max} is defined either by the sensitivity or by a merger frequency, $P(\tau, F = f_{max})$ is the autocorrelattion of a whitened template (self-similarity of a signal under the time shift τ). The word "whitened" means that the template is weighted by the noise amplitude spectral density, $\hat{h}(f)/\sqrt{S_n(f)}$. We impose the normalization condition $P(\tau = 0, F = f_{max}) = 1$. For z we have used a similar definition:

$$z\,(t_0 + \tau, F) = 4\Re \int_{f\,min}^{F} \frac{\tilde{d}(f)\hat{h}^*(f)}{S_n(f)} e^{2\pi i f \tau}\, df. \tag{3.66}$$

First, we show how to recover the χ^2-test described in the previous section using these notations. Take $\tau = 0$ and consider the frequency increment $F_i \to F_i + dF_i$ such that

$$\Delta z_i = \Delta z\,(t_0, 0, F_i + dF_i) - \Delta z\,(t_0, 0, F_i) = z_i - q_i z\,(t_0)\,, \tag{3.67}$$

where as before $z_i = z(t_0, 0, F_i + dF_i) - z(t_0, 0, F_i)$ is the a contribution to SNR from the frequency band $[F_i, F_i + dF_i]$ and $q_i = P(0, F_i + dF_i) - P(0, F_i)$ is the expected fractional contribution (in absence of the noise) to the total SNR. And we recover the χ^2 veto:

$$\chi^2 = \sum_i \frac{(\Delta z_i)^2}{q_i}. \tag{3.68}$$

Next, we will introduce the time domain test used in LIGO data analysis, called autocorrelation test (or veto). The basic idea of this veto is to compare the expected autocorrelation (using whitened template) with what we get from the actual data. Take now $F = f_{max}$ and drop the frequency dependence from the notations:

$$\Delta z_a\,(t_0, \tau_k) = z\,(t_0 + \tau_k) - P\,(\tau_k)\,z\,(t_0)\,, \tag{3.69}$$

similarly we can use the negative lags τ_k. Assuming that t_0 is the true time of signal's arrival, one can easily verify the following noise-average properties:

$$\overline{z(t_0 + \tau_k)} = AP(\tau_k),\quad \overline{\Delta z_a(t_0, \tau_k)} = 0,\quad \overline{\Delta z_a^2(t_0, \tau_k)} = 1 - P^2(\tau_k). \tag{3.70}$$

We can introduce a quantity χ_a which also has χ^2 distribution with N degrees of freedom, where N is a number of lags τ_k:

$$\chi_a^2 = \sum_{k=1}^{N} \frac{\Delta z_a^2(t_0, \tau_k)}{1 - P^2(\tau_k)}. \tag{3.71}$$

We have neglected the possible mismatch between the signal and a template (systematic bias in a model or parameter mismatch).

Now we are in position to generalize this test in time-frequency by introducing

$$\Delta z_{j,\alpha} \equiv \Delta z(t_0, \tau_\alpha, F_j) = z(t_0 + \tau_\alpha, F_j) - P(\tau_\alpha, F_j)z(t_0), \tag{3.72}$$

where we use a time shift τ_α of a template truncated in frequency (up to F_j) for the matched filtering and compare it to the expected result. We can introduce the generalized χ_{t-f}^2 test (veto):

$$\chi_{t-f}^2 = \sum_{j,\alpha} \frac{\Delta z_{j,\alpha}^2}{1 - P^2(\tau_\alpha, F_j)}, \tag{3.73}$$

which distributed as χ^2 with $N_{tot} - 1$ degrees of freedom, where N_{tot} is the total number of time lags and frequency bins (derivation is similar to [13]).

3.5.3 STATISTICAL SIGNIFICANCE OF AN EVENT IN NON-GAUSSIAN NOISE

We want to conclude this section with a short discussion on the estimation of the false alarm probability of a given event candidate. We have already discussed the statistical significance and the false alarm probability in Gaussian noise within the frequentist framework, the problem which arises here is that we cannot model the statistical properties of the noise as it is non-stationary. In this case, we need to use the data itself to infer the noise properties, and for that we need to be sure that the data is free of the GW signal. The LIGO data is noise dominated, and the detectable signals are very rare, so we can shift the data from two detectors by the time lag larger than a light travel time between the observatories (in practice the time lag is significantly larger) and analyze the time-shifted data as a "noise-only." We can be sure that any observed in coincidence signal in the time-shifted data is a noise artifact and we can estimate the distribution of the detection statistic for the noise-generated events. This way we can employ the frequentist method and evaluate the probability of a signal with a given value of detection statistic to be a glitch.

In other GW observations (LISA and PTA), this method does not work and other techniques should be used. The basic principle remains the same: we want to preserve the noise properties (so we have to use the observed data) but eliminate the GW signal from the data, we will say a few more words about this close to the end of this book in Section 4.6.

CHAPTER 4

GW Sources and Observatories

In this chapter we consider the specific aspects of GWs search in three frequency bands: (i) 10–1000 Hz with the ground-based laser interferometric observatories; (ii) 10^{-4}–0.1 Hz with the space-based observatory LISA; and (iii) 10^{-9}–10^{-7} Hz with the Pulsar Timing Array.

4.1   GW LANDSCAPE

In this book, we mentioned several times the LIGO/Virgo GW observatories. Those are the Michelson-based interferometers and we measure the difference in the proper path of the laser photons travelled along two orthogonal arms of equal length. We will give a sketchy derivation of the interaction of an electromagnetic signal with a GW later in this chapter. These detectors are currently sensitive down to 10–20 Hz, below which the seismic noise is rising very steeply. It is possible to push the sensitive region down to a few Hz by going underground (like KAGRA) and cooling the mirrors but the seismic noise prevents us from going below that. The upper bound of the sensitivity is set by a photon shot noise [147].

It is necessary to go to space in order to observe the GW sources in the mHz band. LISA project is approved for a launch as L3 mission in $\sim$ 2034. At the moment of writing this book, LISA has entered the definition phase when the detailed configuration will be finalized, but the basic concept is already fixed and is described below. The LISA consists of three spacecraft forming equilateral triangle trailing the Earth (10–20° behind) on the heliocentric orbit. All spacecraft exchange the laser light (transponding interferometry), and, as in LIGO/Virgo case, we are looking at the relative change in the proper optical path of photons in the dynamical tidal field of GWs. Each spacecraft is shielding the free-floating test mass (a little golden-platinum cube) from the environment. The limitation at the low frequencies come from the residual acceleration (forces) which we cannot eliminate below about 10^{-4} Hz, while at high frequencies (> 0.1 Hz) we are limited by the photon shot noise. The overall position of the sensitive band depends on the armlength: it shifts from low to high frequencies (proportional to the armlength ratio) as we go to lower armlength. We expect that the LISA armlength will be in the range 2–5 mln·km (current configuration has 2.5 mln·km).

In the nanohertz band, 10^{-9}–10^{-7} Hz, we use recycled millisecond pulsars as ultra-stable clocks. The recycled millisecond pulsars are old pulsars which often found in binary systems and were span up by accretion of the matter from its companion. The most attractive feature of those pulsars is stability: any structural deformation, which could cause glitches, have settled down and the period of rotation of these pulsars is very stable. Pulsar Timing Array (PTA) can be seen as multi-arm detector where electromagnetic (e/m) signal travels only in one direction, the main principle (interaction of photons with GWs) is the same as in LIGO/LISA. We measure the deviation in the time of arrival of radio pulses from the expected one: the so-called *timing residuals*. The large part of these deviations could be explained/modeled by various (astro)physical effects, such as: (i) pulsar's spin-down (the energy of the radio emission is extracted from the rotational energy); (ii) relativistic effects affecting a photon propagation in the gravitational field of a pulsar and its companion; (iii) time-dependent dispersion caused by the interstellar medium; and (iv) inaccuracy of the local clock, ephemeris, etc. [95]. Once all these effects are taken into account we are left with tiny residuals which contain the noise (we will discuss it later in this chapter) and the GW signal which is common for all pulsars in the array.

The relative sensitivity of observatories in each of three bands is given in Figure 4.1. For each band we plot the typical sensitivity curve (noise rms needed for detecting monochromatic signal with $SNR = 1$). On top of the sensitivity curves we give typical GW signals expressed in characteristic strain [136], which is $f\tilde{h}(f)$. Such representation is very convenient as it allows easily assess the SNR:

$$SNR = 2\left[\int \frac{h_c^2}{h_n^2} d\log f\right]^{1/2} \equiv 2\left[\int \frac{|f\tilde{h}(f)|^2}{(fS_n(f))}\frac{df}{f}\right]^{1/2}, \qquad (4.1)$$

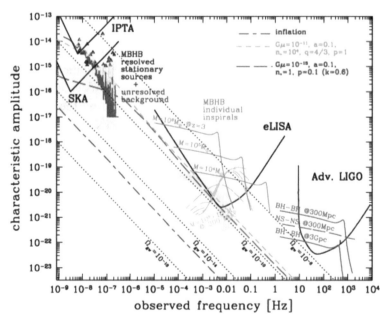

Figure 4.1: GW landscape: characteristic GW strain vs. noise rms in three bands corresponding to the advanced LIGO, LISA, and PTA. Courtesy of A. Sesana.

the ratio between two curves per unit logarithm frequency band (roughly) gives the SNR.

Let us make some simple order of magnitude estimations. Consider coalescence of two equal mass BHs. We approximate the merger time as an instance when the event horizons touch each other. Consider a circular orbit and, for the dominant GW mode with the instantaneous frequency equal to twice of the orbital frequency, we get $f_{GW}^{merg} \approx 1/(\pi 4^{3/2} M) \approx 800$ Hz $(10 M_\odot / M)$. So the stellar origin binary BHs merge in the LIGO/Virgo frequency band (we knew it already), while LISA will see MBH binaries, for instance two BHs with masses $4 \times 10^6 M_\odot$ merge at around 1 mHz. We have chosen this particular mass because it corresponds to an MBH in the nuclei of our own Milky Way Galaxy. We do not observe it directly but we have a good estimate of its mass from the dynamics of the bright stars (so-called S-stars) orbiting around it [69, 71].

In Table 4.1, we summarize the main features of the GW data in three frequency bands observed with different instruments/methods. The transient GWs (i.e., coalescing binaries) are rare in LIGO data, so the signals do not overlap, the signals are weak and one needs to assess the statistical significance of each signal, especially due to non-stationarity of the noise. The signals are in high frequency, so the sampling rate is high and therefore we deal with a large amount of data. In addition to the main data stream, a lot of auxiliary information from monitoring the instrument's subsystems and the environment is recorded and stored. The characteristic wavelength of GW is significantly larger as compared to the arm-length of the detector. We observe the merger of BHs, so we require full GR to describe the GW signal (see Chapter 1).

Table 4.1: Comparison of GW observatories across frequency range

GW Observed	LIGO	LISA	PTA
Frequency (Hz)	20–2,000	10^{-4}–10^{-1}	10^{-9}–10^{-6}
GW rate	Rare and weak, independent of each other	Strong, many overlapping in time and/or frequency	Weak, many, overlap in time and frequency
GW duration	Strongest signals transient, msec. → min.	All the time and transient, hours → years	All the time present in data
GR	Requires full GR	Requires full GR	Leading order
Char. length	$L^{arm} \ll \lambda^{GW}$	$L^{arm} \approx \lambda^{GW}$	$L^{arm} \gg \lambda^{GW}$
Data size	Very large amount sampling 16384 Hz	USB stick sampling 3 Hz	USB stick uneven sampling

The LISA operates in mHz-band and the sampling rate a few Hz is sufficient, the year-worth data is about 1.5 GB, so it can be stored on the USB sticks. The antenna beam function of LISA is rather broad, so it observes the signals coming from different directions all the time, the number of anticipated GW signals is large and some signals could be very strong reaching SNR $\sim$ few $\times$ 1000. We will discuss in more details the GW sources later in this chapter. The data analysis problem which we face with the LISA data is to disentangle and characterize all the GW signals present in the data. Some signals are present in the data all the time, other (like merging BHs) could stay in the band between several hours and several months.

For detecting GWs in the nHz-band we use PTA, and there we rely on the long-term monitoring of stable millisecond pulsars, those radio observations are not regular in time (especially in the past). Currently, we have data which is up to 20 years long but might contain a few month-long gaps. Altogether we have 50–60 pulsars with sufficiently long observation span, and the average cadence of observations is 1/week, so again we deal with a rather small amount of data which could be stored on the USB stick. The prime GW source which we will discuss here is a population of wide MBH binaries in the local Universe. The GW signals from those systems are almost monochromatic and stay in the band all the time. Their superposition forms a stochastic signal at the low-frequency end of the sensitivity band. Another very specific feature of the GW observations with PTA is that characteristic arm-length (which is the distance to a pulsar) is much larger than the GW wavelength, and we will discuss its consequences later in this chapter.

4.2   RESPONSE TO GW SIGNAL

In this section we consider the interaction between an e/m signal and a GW. The detailed derivation can be found in [62, 114]; here we just give few hints and present the final formula. It is convenient to use "TT" gauge introduced in the first chapter (see discussion around Eq. (2.22). In this gauge the coordinates of the emitter (sender) of e/m signal and of the receiver stay the same (not the proper distance though) and we see effect of the interaction of a photon with a GW through the change in its frequency/phase. To derive the response one can use the Killing vectors of the spacetime of the plane GW $\vec{\xi}_1 = \vec{\partial}_\theta, \vec{\xi}_2 = \vec{\partial}_\phi, \vec{\xi}_3 = \vec{\partial}_t + \hat{k}$, where $\hat{k}$ is a unit vector in the direction of GW propagation, and $\vec{\partial}_\theta, \vec{\partial}_\phi$ add to an orthonormal basis. Then we need to consider the null vector connecting sender $s_s^\alpha(t_1)$ and receiver $s_r^\alpha(t_2)$, which is $\Delta s^\alpha = s_r^\alpha - s_s^\alpha$, and take into account that the GW is weak (working in the linear approximation in deviation from Minkowsky space-time). We are interested in expressing Δs^0 component as it reflects the change in the frequency of a photon. The final result is

$$\frac{\Delta \nu}{\nu_0} = \frac{n^i n^j \Delta h_{ij}}{2\left(1 - \hat{k}.\hat{n}\right)}, \tag{4.2}$$

where ν_0 is a frequency of an emitted photon, $\hat{n} \propto \vec{R}_{rec} - \vec{R}_{send}$ is a unit vector in the direction of photon propagation, and

$$\Delta h_{ij} = h_{ij}(t_{send}) - h_{ij}(t_{rec}). \tag{4.3}$$

Denote the Doppler shift introduced by GW as y_{slr}, the subscript reads as the response on the photon emitted by **s**ender and traveling along the path **l**ink to the **r**eceiver: s, l, r

$$y_{slr}^{GW} = \Psi_l \left(t - \hat{k}\vec{R}_s - L_l \right) - \Psi_l \left(t - \hat{k}\vec{R}_r \right), \quad \Psi_l = \frac{n^i n^j h_{ij}}{2 \left(1 - \hat{k} \cdot \hat{n} \right)}, \tag{4.4}$$

where L_l is the distance between receiver and sender and the link is defined by a unit vector $\hat{n} = (\vec{R}_r - \vec{R}_s)/L_l$.

The GW signal can be decomposed in a polarization basis as $h_{ij} = \epsilon_{ij}^+ h_+ + \epsilon_{ij}^\times h_\times$, then

$$\Psi_l = F_+^l h_+ + F_\times^l h_\times, \quad F_{+,\times}^l = \frac{n^i n^j \epsilon_{ij}^{+,\times}}{2 \left(1 - \hat{k} \cdot \hat{n} \right)}. \tag{4.5}$$

Define the signal of the form

$$h_+ = A_+ \cos \Phi(t) = \frac{1}{2} A_+ e^{i\Phi(t)} + c.c. \tag{4.6}$$

$$h_\times = A_\times \sin \Phi(t) = -\frac{i}{2} A_\times e^{i\Phi(t)} + c.c., \tag{4.7}$$

where in the last equality we have used the complex form, $c.c.$ means complex conjugate term. We absorb factors $1/2$ and $-i/2$ into corresponding amplitudes retaining the same notation. Then we have

$$\Psi_l = \left(F_+^l A_+ + F_\times^l A_\times \right) e^{i\Phi(t)}, \tag{4.8}$$

where we have dropped the $c.c.$ part assuming that it is always there. Next, we will make some assumptions which lead to the simplification of the response. We have already mentioned several times that in the matched filtering technique we track very precisely the phase of the signal, while it is somewhat less sensitive to the amplitude. In what follows, we assume that the antenna response and the amplitude of the signal are slowly changing functions of time, namely we assume that

$$F_{+,\times}^l A_{+,\times} \left(t - \hat{k}\vec{R}_s - L_l \right) \approx F_{+,\times}^l A_{+,\times} \left(t - \hat{k}\vec{R}_r \right) \approx F_{+,\times}^l A_{+,\times} (\tilde{t}). \tag{4.9}$$

Indeed, the change in the antenna beam function $F_{+,\times}^l$ is due to the relative motion of the sender and receiver which is usually very slow, more precisely, characteristic time scale of change in the unit vector $\hat{n}$ is much longer than the photon travel time. The variation in the GW amplitude on the same time scale can also be neglected; unless we deal with the PTA, we will come back to this in the PTA section later, and for now assume that it is true (valid for LIGO/Virgo and LISA). We will expand the phase in the Taylor series around a common time t, but before that we decompose the $\vec{R}_s$ and $\vec{R}_r$ as $\vec{R}_{s,r} = \vec{R} + \vec{q}_{s,r}$. For LISA vector $\vec{R}$ will be pointing from the solar system barycenter to the guiding center and for LIGO it points to the beam splitter, while $|\vec{q}_{s,r}| \leq L_l \ll R$. With this notations we can write:

$$y_{slr}^{GW} = \left(F_+^l A_+ + F_\times A_\times\right)_t \left[e^{i\Phi(t-\hat{k}\vec{R}-\hat{k}\vec{q}_s-L_l)} - e^{i\Phi(t-\hat{k}\vec{R}-\hat{k}\vec{q}_r)}\right]$$
$$\approx \left(F_+^l A_+ + F_\times A_\times\right)_t e^{i\Phi(t-\hat{k}\vec{R})}\left[e^{-i\omega(\hat{k}\vec{q}_s+L_l)} - e^{-i\omega\hat{k}\vec{q}_r}\right], \qquad (4.10)$$

where, in the last equality, we have expanded the phase around the common time. We have preserved the term $\hat{k}\vec{R}$ in the phase which is essentially a dominant term in the Doppler modulation of the phase due to relative motion of the GW source and detector. This is a ready-to-use response of a single link (sender-receiver) to a GW wave. In practice, the GW signal is weak and buried in noise, we do not discuss the sources of noise here (referring readers to [147]), however we should mention the dominant noise that is the laser frequency noise (even for the most stable lasers we cannot keep $\delta\nu/\nu_0$ below the expected GW contribution). In order to overcome this problem we use an equal arm Michelson-type interferometer and consider the differential measurement of the response of each arm, which allows us to cancel the laser noise.

We start with considering the ground based interferometers and use LIGO as an example. Each LIGO detector consists of two equal orthogonal arms, where the laser light travels between the beam splitter and the end mirrors.[1] The armlength of LIGO is 4 km and the best sensitivity is achieved at around 100–200 Hz; this implies that $\omega_{GW}L \ll 1$ and we will use this property to further expand the right hand side of Eq. (4.10):

$$e^{-i\omega(\hat{k}\vec{q}_s+L_l)} - e^{-i\omega\hat{k}\vec{q}_r} \approx -i\omega L_l\left(1 - \hat{k}\cdot\hat{n}_l\right) \qquad (4.11)$$

and the single link response for the long-wavelength limit ($\omega L \ll 1$) becomes:

$$y_{slr}^{LIGO} \approx i\omega L_l \frac{n_l^i n_l^j}{2}\left(\epsilon_{ij}^+ A_+ + \epsilon_{ij}^\times A_\times\right)e^{i\Phi(t-\hat{k}\vec{R})}. \qquad (4.12)$$

Consider an equal arm Michelson interferometer schematically presented in Figure 4.2. The beam splitter is located at the point "1", the light departs simultaneously and travels along arm L_2, $1 \to 3 \to 1$ and along arm L_3, $1 \to 2 \to 1$. We use the same index for the arm as for the opposite mirror and the direction of the $\hat{n}_l$ is chosen to be anti-clockwise. Two optical path are shown by red and blue in Figure 4.2, and one can see that if $L_2 = L_3 = L$, the optical path is the same, so the laser frequency noise is canceled if we recombined (subtract) the laser signal at the beam splitter $\Delta t = 2L$ later. The laser noise after the equal round trip is the same while

---

[1]Here we use an over-simplified picture of the LIGO to grasp the main physics, in practice LIGO uses several cavities to increase the sensitivity of the instrument [147].

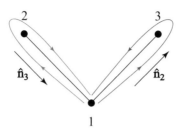

Figure 4.2: Equal arm Michelson.

the response to the GW depends on the relative orientation of source location $(-\hat{k})$ and the photon's path. Let us explain the example of equal arm interferometer in more details, so the unequal arm case can be easily derived by a reader. Consider a single link consistent of a sender (s) and receiver (r). We neglect all other noise sources and for the laser noise we have:

$$y_{slr}^{l.n.} = p_s(t - L_l) - p_r(t), \tag{4.13}$$

where the sup-script $l.n.$ stands for "laser noise." Consider the blue path. The first part where the laser photons traveled from "1" to "2" is given as $y_{1-32}(t + L_{-3})$, where $y_{slr} = y_{slr}^{GW} + y_{slr}^{l.n.}$, L_{-3} means that we travel along the arm L_3 but in the direction opposite to $\hat{n}_3$ which is important if the armlength depends on time, and t is measured by the beam splitter ("1"), $t = 0$ being the time of emission. To conclude the round trip, the laser photons travel from "2" back to "1" and described as $y_{231}(t + L_{-3} + L_3)$, so the blue path is given by $y_{1-32}(t + L_{-3}) + y_{231}(t + L_{-3} + L_3)$. Similarly, we can get expression for the red path: $y_{123}(t + L_2) + y_{3-21}(t + L_2 + L_{-2})$. More frequently used notations define zero time as a time of reception (or recombination) of the laser photons at the beam splitter (not the time of emission), so we will re-write our expressions using this convention: $(t + L_{-3} + L_3) = (t + L_2 + L_{-2}) \rightarrow t$. Substitute the laser noise contribution to the each path:

$$1 \rightarrow 3 \rightarrow 1 : \quad y_{3-21}^{l.n.}(t) + y_{123}^{l.n.}(t - L_{-2}) = p_1(t - L_2 - L_{-2}) - p_1(t)$$
$$1 \rightarrow 2 \rightarrow 1 : \quad y_{231}^{l.n.}(t) + y_{1-32}^{l.n.}(t - L_3) = p_1(t - L_3 - L_{-3}) - p_1(t). \tag{4.14}$$

One can easily see that, if the arms are equal, the differential measurement cancels the laser noise and we have

$$X_{equal-arms} = y_{3-21} + y_{123,-2} - (y_{231} + y_{1-32,3}) = y_{3-21}^{GW} + y_{123,-2}^{GW} - (y_{231}^{GW} + y_{1-32,3}^{GW}), \tag{4.15}$$

where we have introduced a short-hand notation for the time delay operator, $y_{123,-2} \equiv y_{123}(t - L_2)$. Substituting the single-link response (4.11) into (4.15) we get

$$X^{LIGO} = i\omega L(F_+ A_+ + F_\times A_\times)e^{i\Phi(t + \hat{k}\vec{R})}, \tag{4.16}$$

where $F_{+,\times}$ is the antenna beam function which we have already seen in previous chapter:

$$F_{+,\times} = \left(n_2^i n_2^j - n_3^i n_3^j\right)\epsilon_{ij}^{+,\times}. \tag{4.17}$$

Note that what is actually measured by LIGO is the phase difference (not the frequency difference given above). We can integrate the fractional frequency to get $\Delta\phi_X$ which absorbs the factor $i\omega$. We have reproduced the LIGO response (see Eq. 3.18); it is proportional to the difference in projections of the GW signal h_{ij} on the arms of the detector and, as we have already seen, the sensitivity scales with the armlength L. We should also say few words about the term $\hat{k}\cdot\vec{R}$ in the phase. This term is important for the long-lived GW signals, namely if the duration of the GW signal in the LIGO band is comparable to the scale at which $\vec{R}(t)$ changes appreciably. A good example is (almost) monochromatic signal coming from the slightly deformed neutron star; the GW signal lasts the entire duration of the observation and we need to take into account modulation of the signal due to annual motion of Earth around the Sun. For the short transient signals, like those from binary systems, which last less than a minute, this term is constant and we can completely neglect it. Note that this term contains $\hat{k}$ which is essentially the sky location of a GW source, and if the contribution of this term is measurable, it allows us to localize the GW source. Last comment before we move to LISA is that, in general, the antenna beam function $F_{+,\times}$ is a function of time and is responsible for the amplitude modulation of the signal, however, it is important only for the long-lived signals and the time dependence could be neglected for the short-duration transient sources which are the main subject of this book.

Now we turn our attention to LISA. The exact details of the LISA configuration are yet to be finalized, however, the main concept remains unchanged already for a long time. LISA consists of three identical satellites in the free fall on the heliocentric orbit. The orbit of each spacecraft is chosen in such a way as to form an equilateral constellation at any time. The distance between the satellites is not fully defined, but, at the moment of writing this book, the armlength is 2.5 mln·km. The spacecraft exchange the laser light with each other to form a transponding interferometry (the light received by a partner spacecraft is not reflected but amplified and transmitted as a fresh high-power beam that is phase-locked to the incoming weak beam, with a fixed offset frequency). Effectively we can split the measurements into three Michelson-type interferometers as shown in Figure 4.3 where each spacecraft represented by a bullet is enumerated and solid lines show the laser links. Note, however, that those three interferometers are not independent as they share the same links. We can form other three combinations of measurements with an independent noise (referred to as A, E, T [138]), where A, E have a similar sensitivity and T combination has comparable sensitivity only for the high frequencies such that $f > f_*$, $f_* L = 1$ (where the gravitational wavelength is comparable to the armlength). At low frequencies the response to the GWs for the T combination is significantly suppressed and it can serve as the instrumental diagnostic channel (considering it to be instrumental noise dominated).

The biggest difference between LIGO and LISA is that the LISA's arms are not equal, moreover, they are slowly changing with time. Three spacecraft are in the free fall around the sun

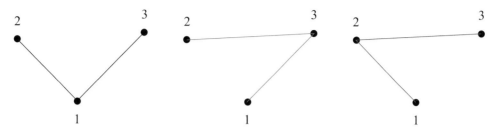

Figure 4.3: Three effective, correlated Michelson configurations. Each spacecraft is enumerated and represented by a dot.

and the orbit is sufficiently stable to keep the arms *almost* equal. We can create a combination of measurements for unequal arm configuration similar to 4.2, and call it unequal-arm Michelson combination:

$$X = y_{1-32,32-2} + y_{231,2-2} + y_{123,-2} + y_{3-21} - [y_{123,-2-33} + y_{3-21,-33} + y_{1-32,3} + y_{231}].$$
(4.18)

In this expression the term in the brackets (negative) describes the red path and the positive terms correspond to the blue path shown in Figure 4.4. The main idea is the same as for the equal arm configuration, by applying the appropriate delays to the measurements we track the propagation of the laser frequency noise so that we have equal optical path traveled in each direction. This optical path is shown as the blue and red path in Figure 4.4, and by subtracting one from another we completely eliminate the laser frequency noise, while the response to GW signal is different for the laser light travelled in the red and blue path. The response to each link is well approximated by expression (4.10). Schematically the LISA constellation is presented in Figure 4.5.

The motion of each spacecraft can be seen as a heliocentric motion of the guiding center O plus rotation of the spacecraft (with period one year) around O. The constellation is inclined 60° to the ecliptic and schematically shown in Figure 4.6. We use anti-clockwise direction for a

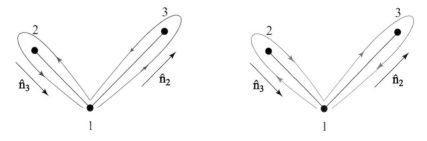

Figure 4.4: Unequal arm Michelson.

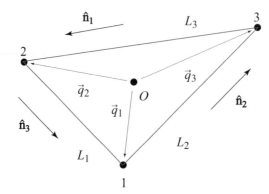

Figure 4.5: Schematic representation of the LISA constellation. The bullet points represent the position of each spacecraft. We use anti-clockwise propagation as positive, and the arms are labeled as opposite spacecraft.

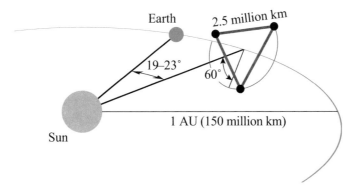

Figure 4.6: LISA's orbit. Each spacecraft is in the free-fall forming equilateral triangle, trailing behind Earth. LISA's plane is inclined 60° to the ecliptic.

positive links $\hat{n}_i$, and the arm opposite to a given spacecraft has the same index. It is convenient to work in the solar barycentric reference frame and the position vector of each spacecraft is split into a sum of the position of the guiding center O (at the distance of 1 astronomical unit, or 499 secs in geometric units) and the vectors $\vec{q}_i$ introduced earlier.

The measurement along the link l can be presented (in somewhat simplified way, keeping only major contributions) as

$$s_{sr} = y_{slr}^{GW} + b_{sr} + (p_{s,l} - p_r) + ((\vec{a}_s \hat{n}_l)_{,l} - (\vec{a}_r \hat{n}_l)), \qquad (4.19)$$

where y_{slr}^{GW} is our main measurement (response to the GW), b_{sr} is the contribution of the laser shot noise in the link between sender **s** and receiver **r**, the blue term is the contribution of the laser frequency noise p, and the red term is the residual acceleration ($\vec{a}_i$) noise. We have used here again the introduced above notation for the delay operator $p_{s,l}(t) \equiv p_s(t - L_l)$. One can check by explicit substitution of this expression into (4.18) that the laser noise cancels. Note that the response to GW in X is proportional to the second derivative of the strain, $\ddot{h}$. The first derivative comes from the fact that we measure the fractional change in the frequency of the laser light, and the extra derivative is added due to extra loop required for cancelling the laser frequency noise (see Figure 4.4).

Now we turn our attention to the last term (in square brackets) in the response to GW:

$$y_{slr} = \left(F_+^l A_+ + F_\times A_\times \right)_t e^{i\Phi(t-\hat{k}\vec{R})} \left[e^{-i\omega(\hat{k}\vec{q}_s + L_l)} - e^{-i\omega\hat{k}\vec{q}_r} \right]. \qquad (4.20)$$

The vectors $q_i \sim L_l \sim L$ are all of the order of the armlength, so this term becomes important where $\omega L \sim 1$ which corresponds to 20 mHz for the armlength $L = 2.5$ mln·km. At the frequencies which much lower than that we can use (approximately) the long-wavelength approximation similar to LIGO, however it becomes increasingly inaccurate as we approach 20 mHz, so it is advised to go beyond the leading order expansion in ωL. Starting from about 20 mHz the constellation is sensing the delays in the propagation of GW between spacecraft and that is what is encoded in the last term (square brackets) in Eq. (4.20). This is important point and we want to elaborate it a bit further. We can construct three Michelson-like interferometers shown in Figure 4.3, which we call X, Y, Z and they can be obtained from (4.18) by permutation of indices $1 \to 2 \to 3 \to 1$. We have already mentioned that they are not independent as they share a link (pair-wise). At the low frequencies where we can employ the long-wavelength approximation, we can construct the so-called *null* stream where the gravitational signal is significantly suppressed. Indeed, at low frequencies the response function is almost LIGO-like and we can express two polarization states h_+, $h_\times$ in terms of two Michelson measurement and eliminate the GW signal from the third stream [138]. In general, out of three noise-correlated Michelson measurements we can construct three streams with uncorrelated noise (assuming that the noise in each link is uncorrelated and has similar statistical properties), we call them A, E, T [111], and those are

$$A = \frac{Z - X}{\sqrt{2}}, \quad E = \frac{X - 2Y + Z}{\sqrt{6}}, \quad T = \frac{X + Y + Z}{\sqrt{3}}. \qquad (4.21)$$

Those combinations are not uniquely defined, and the numerical coefficients were chosen so that the noise power spectral density of A and E streams are equal $S_A = S_E$. The third stream T is the one where the GW signal is significantly suppressed at low frequencies, however it has sensitivity comparable to A, E at high frequencies. This is an explicit manifestation of the response function at high frequency, we sense the time delay in propagation of GW between spacecraft. Note that we still can construct the null stream at high frequency but for a particular direction on the sky [102].

Mathematically, the null stream comes as yet another solution of the TDI combination which removes the laser frequency noise which is called Sagnac (fully symmetric) combination:

$$\zeta = y_{312,2} - y_{2-13,3} + y_{123,3} - y_{3-21,1} + y_{231,1} - y_{1-32,2}, \qquad (4.22)$$

where the positive terms correspond to the anti-clockwise propagation of the laser frequency noise along the arms and the negative terms are for the clockwise (full circle) propagation.

The null stream is important to characterize the instrumental noise, since the response function has significantly suppressed the GW signal, whatever is left there will be associated with the noise (possibly with some non-stationary features). This is a very desired data combination especially in the signal dominated data stream like LISA is expected to be. However, we should emphasize that the null stream is a combination of several measurements with delays, therefore, it is not a trivial task to predict how the observed instrumental glitch will appear in A, E streams, because the response is different.

In this section we have given a general expression for the response of the electromagnetic signal propagating in the tidal field of a gravitational wave given by Eq. (4.2). We have shown how to derive the response of the LIGO detector which operates in the long-wavelength limit (the gravitational wavelength is much larger than the detector's arm). The LISA's spacecraft are shielding free-falling test masses, so the length of arms varies slightly in time, and we need to employ time-delay-interferometry technique to cancel the laser frequency noise. In addition, we have shown that at some frequency the gravitational wavelength becomes comparable (equal) to the LISA's arm, which makes the response function more complex as it is a function of time and frequency. The Michelson unequal arm combination of measurements is given by Eq. (4.18). We still have to discuss Pulsar Timing Array and this will be the topic of the next section.

4.3   SEARCHING FOR GRAVITATIONAL WAVE SIGNALS WITH PULSAR TIMING ARRAY

We have decided to separate Pulsar Timing Array to a separate section. The reason for that is twofold (i) it operates completely in the short-wavelength limit $L \gg \lambda^{GW}$ and (ii) the source of an electromagnetic signal is provided by nature. We use millisecond pulsars as ultra-precise clocks. In this book we will give only a brief overview of pulsars and refer readers to a very nice book [95] for more information. Pulsars are the neutron stars which are remnants of the massive $M \geq 7M_\odot$ stars which have undergone the supernovae explosion and were not heavy enough to form a BH. In particular, we are interested in the millisecond radio pulsars. Those pulsars were born in binary systems, and being heavier, evolve faster than their companion. The newly born neutron star is often seen as a pulsar, where the beamed radiation coming from the magnetic polar caps (which are misaligned with the rotation axis) sweeps across the line of sight and we see the set of regularly appearing pulses. This is very similar to the light coming from the lighthouse. The period of rotation (spin) of young neutron stars varies quite a lot from a few Hz to a few tens of Hz (30 Hz for Crab pulsar and 11 Hz for Vela pulsar), but it is somewhat unstable, the pulsars *glitch*, within a short time the spin frequency increases by $\delta f_{spin}/f_{spin} \sim 10^{-7} - 10^{-6}$ and then it settles down. Those glitches are not completely understood but usually associated with the interior of the neutron star (most likely they are caused by a differential rotation and stresses produced by the superfluid vortices which crack the outer layer (crust)). The basic theory behind the radio pulsars is that the rotational energy is transformed into the magnetic dipole radiation causing an increase in the spin period. The situation is somewhat different if the pulsar is in the binary system. The companion, after burning all hydrogen in the core, moves off the main sequence along red-giant-branch, fills up its Roche lobe, and starts accretion of gas on the pulsar. The accreting material transfers the angular momentum to the pulsar and spins it up. At the end we have, what is called, recycled old millisecond pulsar, where the "millisecond" refers to its spin period (typically several milliseconds). Those millisecond pulsars are very stable rotators, the interior has already settled down, and usually they do not glitch. Some of the observed millisecond pulsars are so stable that they beat the terrestrial time standards on the long (decades) time scale, and it is exactly this feature which we will exploit.

The radio-telescopes around the globe monitor regularly the most stable millisecond pulsars. The profiles of individual pulses have a lot of microstructure, but the profile averaged over about an hour, is quite stable and it is used to measure the time of arrival (TOA) of a radio pulse. We know very well the spin of a pulsar, so we can predict when the next pulse should arrive, however there are small deviations from the expected time of arrival; those deviations are called *residuals*:

$$\delta t = t_{toa}^{p} - t_{toa}^{o}, \tag{4.23}$$

where t_{toa}^{p} is the predicted time of arrival and t_{toa}^{o} is the measured time of arrival. The accuracy with which we measure the TOA is inversely proportional to the strength of the radio pulses and depends on the stability/accuracy of the estimated profile. The "brightness" of pulsars depends on the distance and on the sensitivity of a radio-telescope. The largest part of the deviation in TOA from the predicted could be explained by various physical process:

$$t_{toa} = t_{toa}(P, \dot{P}, \ddot{P}, \Delta_{clock}, \Delta_{dm}, \Delta_{SSB}, \Delta_E, \Delta_S), \tag{4.24}$$

where we need to take into account (i) the fact that the spin of a pulsar P is decreasing, by considering the first and second derivatives of the rotational period, $(\dot{P}, \ddot{P})$, (ii) the error in the local clock, $\Delta clock$ (as compared to the terrestrial standards), (iii) the delay, Δ_{dm} related to the dispersion due to propagation of the radio signal in the interstellar medium, (iv) the transformation from the solar system barycenter to the detector frame, Δ_{SSB}, (v) the time delay (Einstein term, Δ_E) due to motion of the pulsar and gravitational redshift, and (vi) the Shapiro delay, Δ_S, which comes from the propagation of the radio signal in the curved space-time of the pulsars' companion. The described above timing model parameters ($\vec{\lambda}_{tm}$) are obtained from the fit which minimizes the residuals. Note that the error in the pulsar's sky location and/or in the proper motion lead to the periodic (with period 1 year) oscillations in the residuals, therefore we cannot measure the GW signal with period 1 year (and its multiple), it is removed by the fitting procedure. After applying the timing model to the TOAs we have

$$\delta t = t_{toa}^P\left(\vec{\lambda}_{tm}\right) - t_{toa}^o = \delta t_{err} + \delta\tau_{GW} + \delta t_n, \tag{4.25}$$

where δt_{err} is due to possible (small) error in the timing model parameters $(\vec{\lambda}_{tm})$, δt_n is the measurement and the pulsar noise, and, finally, $\delta\tau_{GW}$ is due to propagation of the radio signal in the time dependent tidal field of a GW. Here we will not talk much about the errors in the timing model, in the data analysis we can rather easily marginalize the results over those uncertainties [139, 141]. The measurement errors might not be complete as they might lack the systematic errors intrinsic to each measuring back-end (and they might also be different for each pulsar), therefore the data analysis techniques should take into account possible biases and marginalize over them. There is a "red noise" (noise rising at low frequencies as $f^{-\gamma}$ in the power spectral density of TOAs) coming from pulsars themselves, this noise corresponds to stochastic fluctuations in the spin of a pulsar. Other main sources of noise and systematic biases come from inaccuracy in the solar system ephemerids and variations in the dispersion measurement due to relative motion of a pulsar and the radiotelescope.

Let us consider the response to the plane GW signal τ_{GW}. We can see pulsar–radio-telescope as a single arm (sender-receiver with only a one-way link) of a detector and we considering the deviation not in the frequency of arriving pulses but the deviation in the phase:

$$r(t) = \int_0^t \frac{\delta\nu}{\nu}(t')dt' \tag{4.26}$$

$$\frac{\delta\nu}{\nu} = \frac{1}{2}\frac{\hat{p}^i\hat{p}^j}{1+\hat{p}.\hat{k}}\Delta h_{ij} \tag{4.27}$$

where the unit vectors $\hat{p}$ and $\hat{k}$ denote the direction to the pulsar and propagation of the GW as shown in Figure 4.7.

We have also defined

$$\Delta h_{ij} = h_{ij}(t_p) - h_{ij}(t = t_e) \tag{4.28}$$

which is the difference between GW strain at the moment of emission (t_p) and at the moment of reception $t = t_e$ ("p" stands for pulsar and "e" stands for Earth):

$$t_p = t - \tau_\alpha. \tag{4.29}$$

The first term $h_{ij}(t_p)$ is called the *pulsar* term and second is the *earth* term, the contribution of GW in between emission and reception averages out. Note that the difference between pulsar and earth time, τ_α, is different for each pulsar, the greek letter are indexing the pulsars (α, β):

$$\tau_\alpha = L_\alpha\left(1 + \hat{k}\cdot\hat{p}_\alpha\right). \tag{4.30}$$

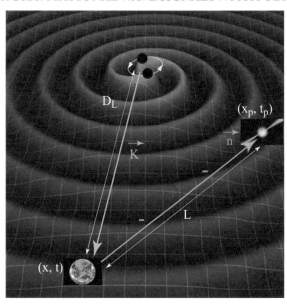

Figure 4.7: Image showing GW source–pulsar–Earth system.

Let us discuss a bit Eq. (4.30). First, note that if $\hat{k} \cdot \hat{p}_\alpha = -1$ the response is exactly zero. This is sometimes referred to as "surfing the wave." Second, it depends on the distance to the pulsar L_α which is not very accurately measured, 10–40% accuracy. The typical distance to the pulsar is few kpc, which corresponds to a delay $\tau_\alpha = 1.1 \times 10^{11} \sec \left[\frac{L_\alpha}{1\text{kpc}} \right] (1 + \hat{k}.\hat{p}_\alpha)$ few hundreds to few thousands years. The PTA is sensitive to the GW in the nano-Hertz band ($\sim 10^{-9} - 10^{-7}$ Hz), so the PTA is completely in the regime where $\omega L \gg 1$ (short-wavelength limit).

We will discuss in details the GW sources in the next section, however, let us briefly mention a few sources here. The most promising source (as usual) are binary systems, but the GW amplitude should be rather large to be detectable with PTA (as one can see in Figure 4.1). This implies that the system has to be very heavy and somewhat nearby, the frequency tells us that the system is relatively broad. And indeed the binary to be observed by PTA should consist of two supermassive BHs with masses 10^7–$10^9 M_\odot$ with an orbital period between a fraction of a year and a dozen of years (or so). We will come back to this when we discuss the sensitivity curve for PTA. For now we assume a binary of supermassive BHs in a quasi-circular orbit and we compute how much the binary has evolved over τ_α. The change in the GW frequency over that time interval is

$$\Delta f_\alpha^{GW} = \int_{t-\tau_\alpha}^{t} \frac{df}{dt'} dt' \approx 15 \text{ nHz} \left(\frac{M_c}{10^{8.5} M_\odot} \right)^{5/3} \left(\frac{f}{50\text{nHz}} \right)^{11/3} \left(\frac{\tau_\alpha}{1\text{kpc}} \right), \qquad (4.31)$$

where M_c is a chirp mass. Note that it is rather a strong function of the GW frequency $f(t)$ of the earth term. The difference Δf_α^{GW} between GW frequency of the pulsar and the earth terms for this particular system is measurable, the Fourier frequency bin is of the size $df = 1/T_{obs} \approx 3$ nHz for the 10-years-long millisecond pulsar monitoring program. There are pulsars in the array which were observed for about 20 years. At the same time, the change in the GW frequency over the observation time is completely negligible, indeed we can use again the above formula and notice that 1 kpc is 4200 (light)years which we need to compare to 20 years of observation. The GW frequency of pulsar and earth terms can be well separated and appear as two monochromatic signals in residuals of each pulsar for systems with a high chirp mass and at high frequencies. Note also that the earth-term will have the same frequency and it is fully coherent across all pulsars whereas the pulsar term appears at frequency individual for each pulsar.

Let us compute the response of a single pulsar to a plane monochromatic gravitational wave propagating in the direction $\hat{k}$ and we start with the earth term. The GW signal can be written as

$$h_{ij}(t) = \left(\epsilon_{ij}^+ A_+ + \epsilon_{ij}^\times A_\times \right) e^{i\omega_{gw}t} \tag{4.32}$$

substituting this form into the response expression (4.27) we obtain

$$r_E = \frac{A}{\omega_{gw}} \left[F_+^\alpha \cos\left(\omega_{gw}t + \phi_0\right) + F_\times^\alpha \sin\left(\omega_{gw}t + \phi_0\right) \right]. \tag{4.33}$$

The antenna response for a single pulsar-Earth link is a function of the relative position of GW signal and a pulsar on the sky:

$$F_{+,\times}^\alpha = \frac{1}{2} \frac{\hat{p}_\alpha^i \hat{p}_\alpha^j}{1 + \hat{p}_\alpha \cdot \hat{k}} \epsilon_{ij}^{+,\times}. \tag{4.34}$$

If we have many pulsars (many arms) we can determine the sky position of the GW signal by measuring the $F_{+,\times}^\alpha$: we need $3 \times N_{gw}$ roughly equivalent and not co-located pulsars in order to localize N_{gw} GW sources [25] using the earth-term only.

We can derive similarly the pulsar term:

$$r_P^\alpha = \frac{A_\alpha}{\omega_{gw}^\alpha} \left[F_+^\alpha \cos\left(\omega_{gw}^\alpha t + \phi_0^\alpha\right) + F_\times^\alpha \sin\left(\omega_{gw}^\alpha t + \phi_0^\alpha\right) \right]. \tag{4.35}$$

In general, the amplitude A_α, frequency ω_α, and phase ϕ_0^α are different for each pulsar. They are not all independent, in principle, they are functions of the distance to the pulsar L_α and the chirp mass (M_c). We see that the inclusion of the pulsar term seriously increases the number of independent parameters (by $N_{psr} + 1$, where N_{psr} is a number of pulsars in the array). The total response is $r = r_P - r_E$.

Note that for the circular binary $A \propto \omega^{2/3}$ so the overall amplitude is proportional to $\omega^{-1/3}$, and, since $\omega_{gw}^\alpha \le \omega_{gw}$, the amplitude of the pulsar term is higher. This factor, $\omega^{-1/3}$, also defines the slope in the sensitivity curve which is rising for the high frequencies as $\omega^{1/3}$ (if we plot GW strain which produces SNR~ 1). The low-frequency bound in the sensitivity is determined by the length of observations, and it is usually not as sharp as in the Figure 4.1 because different pulsars have different sensitivity (accuracy in timing) and were observed for a different duration.

At low frequencies we could expect that the difference between the frequency of the pulsar and earth terms is not measurable; in other words, those frequencies fall within the same Fourier bin (for example you can try $f = 5$ nHz in the expression (4.31)). In this case, the pulsar term completely messes up the coherence of the earth-term, and we deal with a single monochromatic signal in each pulsar's data.

One important peculiarity of the PTA data is that it is not evenly sampled. The radio observations for the past few years are quite regular, but we cannot say the same about early days ($\sim$ 10 years ago) the data contains irregular gaps sometimes few months long. This makes it easier to work with the data in the time domain. One could perform an approximate Fourier transformation using for example Lomb-Scargle method (based on the least square fit), but in the end it does not help much due to error (uncertainty) associated with this method.

4.4   GRAVITATIONAL WAVE SOURCES IN LIGO BAND

In this section we consider the main GW sources in the kilo-Hertz frequency band.

4.4.1   DETECTION OF MERGING BLACK HOLE BINARIES WITH LIGO DETECTORS

We have focused our discussion in this book around GWs from binary systems, and it was done deliberately. The first gravitational wave signal detected by LIGO and VIRGO collaborations was from two coalescing black holes. Soon after there were two more confirmed detections of GWs from binary black holes. There was one more event which had lower statistical significance (simply due to being further away and, as a result, with lower SNR) but according to the preformed analysis also corresponds to the merging binary black hole systems.

Those are results of the first observing run "O1" and beginning of the second "O2," since then many binaries were detected but we will concentrate on those four (we will comment on some other detected events at the end of this section). We briefly characterize each of these events. Figure 4.8 shows all four observed binaries; the size of each blue circle is proportional to the black hole's mass. The order of events corresponds to its discovery time (from the left to the right).

The very first event GW150914 was detected (as its name says) on September 14, 2015 [4–6]. It was the heavy system with masses $m_1 = 36 \pm 4 \, M_\odot$ and $m_2 = 29 \pm 4 \, M_\odot$. The SNR of this event was ≈ 24 and its duration in the LIGO band (30–250 Hz) was ≈ 0.2 sec. Since it was a heavy system, we observed only the late inspiral followed by a merger and ringdown. The distance to the source was estimated to be $D_L = 410 \pm 160$ Mpc (which corresponds to

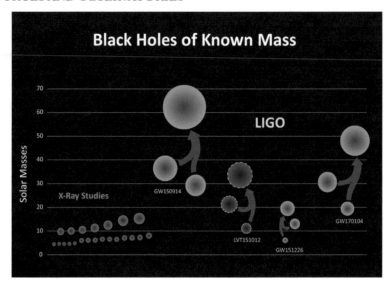

Figure 4.8: Chart of observed black hole masses. The purple circles are the estimates from the X-ray binaries, the blue circles are from the GW observations. The size is proportional to the mass.

the redshift $z \approx 0.1$). The residual BH has a mass of about $62 \pm 4 \, M_\odot$, which implies that the energy release in GWs was roughly corresponding to $3 \, M_\odot$. The statistical significance of this event is encoded in the associated false alarm rate, which was $\approx 6 \times 10^{-7} \, \mathrm{yr}^{-1}$, that means that similar event could be generated by noise only once in 1.6 mln·years. The biggest surprise with this source was the BH masses, all previous estimates for the BH candidates (purple circles in Figure 4.8) coming from the observation of the X-ray binaries where $\lesssim 15 \, M_\odot$. Formation of such BHs requires the large mass progenitor stars with low metallicity (almost pure hydrogen with very few heavier elements), similar to the composition of the very first population of stars (pop-III stars). There are several channels for formation binary BH systems: (i) both BHs are born as a result of supernovae explosion after evolution of the massive binary star system and (ii) BH binary is formed from single BHs in the dense environment (like a globular cluster) by dynamical capture or by exchanging the companions in a mixed binary systems. We will speak about the BH' spins a bit later. Based on those four detections and using the results from the first observing run as a prior, the inferred binary BH merger rate in the local Universe is 12–213 $\mathrm{Gpc}^{-3} \, \mathrm{yr}^{-1}$ [10].

The second event, GW151012 [2], had significantly lower SNR (SNR ≈ 9.7) and the corresponding false alarm rate is 1 every 3 years. This source was significantly further away $D_L \approx 1$ Gpc (redshift $z = 0.2 \pm 0.09$). Because of the low SNR, the estimation of parameters is significantly worse; nevertheless, the masses of BHs are $m_1 = 23 \pm 10 \, M_\odot$, $m_2 = 13 \pm 5 \, M_\odot$ which is lower than in the first event.

The third GW event happened on Boxing Day (the day after Christmas), that is GW151226 [3]. The statistical significance of this event was similar to the first one, and SNR ≈ 13. This is the lightest binary BH system among four: $m_1 = 14 \pm 4 \, M_\odot$, $m_2 = 7.5 \pm 2 \, M_\odot$, and, as the result, the duration of the signal was about 1 sec (in the frequency range 35–750 Hz). The main contribution to the SNR comes from the inspiral part of the signal and this was reflected in the estimation of parameters, especially banana-like 2-D posterior distribution function in component masses [3]. This is a typical feature of the inspiral, where we can measure very well only the chirp mass which defines this particular shape. Observation of a merger allows us to measure well the total mass of the system, which breaks the strong mass degeneracy of the inspiral part (as it was the case for the GW150914). The distance to the source was roughly the same as for the first event $D_L = 440 \pm 190$ Mpc. The in-band duration of this signal is also the reason to have similar false alarm rate between the "first" and the "boxing-day" events despite that the later has half SNR of the former: it is harder to produce pure noise generated a chirp-like event with a long duration.

The fourth GW event was detected about a month after the beginning of the second observational LIGO' run (O2), that is GW170104 [10]. It is relatively heavy binary with masses $m_1 = 31 \pm 7 \, M_\odot$, $m_2 = 19 \pm 5 \, M_\odot$, closing the gap between the first and second events. As a result of its rather heavy total mass, the duration of the signal was ≈ 0.3 sec (between 30 Hz and 370 Hz), with the merger-ringdown part of the signal dominating over the inspiral. The SNR of this event was ≈ 13 and, as a result, the false alarm rate was once per 70,000 years. This source was almost twice further away as compared to the first and the third events, $D_L = 880 \pm 400$ Mpc (redshift $z \approx 0.2$).

It is time to discuss the spins of BHs in the observed binary systems. First, we have to mention the observational selection effect: we preferably see the systems which are "face-on" or "face-off." Indeed, if you look at the expression (2.59) you see that the magnitude of the GW is maximum (provided we have fixed the distance to the source) if the inclination angle is 0 or π. This makes somewhat difficult to identify precessing binaries. The precession of the orbital plane about the total momentum of the system is the result of the spin-orbital coupling, which we have already mentioned in Section 2.4.2. During the precession the orbital angular momentum and spins rotate around the total momentum of the system. Such precession is clearly visible/identifiable for the systems seen "edge-on," and less clear if the system is "face-on/off." The observed systems seem to be close to the "face-on/off" configuration, in addition, the spins enter the GW phase starting from 1.5 PN order, so their contribution is generically weak. All these couples to a rather low SNR (< 25) of all observed systems and makes it hard to estimate very well the spin magnitudes and orientation.

In Figure 4.9 we summarize the spin information extracted from the GW signals using Bayesian techniques described in Section 3.4. We project spins of primary (more heavy) and

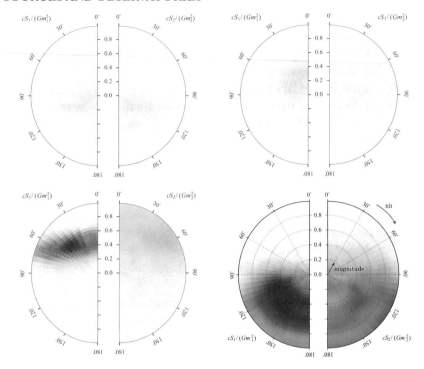

Figure 4.9: Projection of BH spins on the plane orthogonal to the orbital plane. These plots show 2-D distribution of the spin magnitude (as the distance form the center) and the tilt angle (angle between the spin and the orbital angular momentum). Zero angle corresponds to the spin orthogonal to the orbital plane and aligned with the orbital angular momentum. The top panel: GW150914 (left) and GW151012 (right); the low panel: GW151226 (left), GW170104 (right).

secondary BHs on the plane orthogonal to the orbital plane, so the orbital angular momentum is pointing upward (to 0°). The spin of a primary BH is always shown in the left semi-circle. We show the two-dimensional distribution of the amplitude of spin and its tilt angle, the angle between the spin and the orbital angular momentum, $\cos \theta_i = (\hat{L} \cdot \hat{S}_i)$. The spins (anti)aligned with the orbital angular momentum corresponds to (180°) 0°, and the orbital plane is at 90°. The depth of color corresponds to the posterior probability (per pixel). We cannot say much about the spins of the GW150914 and GW151012, most likely the spins were not very high, the 90% credible interval tells us that $|\vec{S}_1| < 0.7m_1^2$ and the secondary's spin is very poorly constrained for both events. The third and fourth events are more interesting, we can put lower bound on the primary spin for GW151226, that can be clearly seen from the figure. The 90% credible interval for the "boxing day" event is $0.8m_1^2 > |\vec{S}_1| > 0.2m_1^2$, which is very exciting as it excludes the completely non-spinning configuration. Unfortunately, we cannot make a strong statement about the primary tilt angle, there is an indication that the spin is misaligned, but the aligned configuration has also significant support in the posterior distribution. The interesting feature of the fourth merger is that the tilt angle seems to be rather in the southern quarter (tilt angle > 90°) which implies that the primary BH spins in the direction opposite to the orbital rotation. Note that those projections should be referred to a particular time, and those plots correspond to the instantaneous GW frequency 30 Hz, however, if there is precession, then we would expect those disk plots to evolve in time.

Now we turn our attention to the fundamental physics, namely testing GR with the detected GW signals. Here we will only briefly mention the performed test and we refer the reader to [6, 10] and references therein for a more detailed description. The first test is called *parameterized test of GR*, the basic idea is that the phase of GW is described as an expansion in small parameter v/c and GR predicts specific numerical parameters in front of each term, so we can introduce a phenomenological modification of each numerical parameter and infer its value (or probability distribution function) from the observed data. This test could be extended beyond the inspiral part (described by that expansion) to merger and ringdown in a similar way. The second test is a consistency test between the inspiral and merger parts of the observed GW signal. We artificially split the analysis into two parts: we analyze independently inspiral but cutting off the search template at some frequency and the post-inspiral part. The inferred parameters could be mapped onto the (final) spin and the mass of the remnant BH assuming GR relationship, and we compare the results obtained from two independent analysis. The GR has passed this consistency test if the two-dimensional posterior pdfs from two analysis have significant overlap. Next, we test the propagation properties of the GW signal. The GR tells us that the graviton is a massless particle propagating with the speed of light. If graviton, however, has a non-zero mass then it becomes sub-luminal and we expect the dispersion of the GW signal (low frequencies propagate slower than the high frequencies). In addition, various theories predicting the Lorentz invariance violation lead also to the modification of the dispersion but the propagation could also be super-luminal. Finally, we could subtract the GR template (model) best matching the observed signal from the data and test consistency of the residuals with the noise. All performed test show that the GR theory can explain all currently observed GW events. In addition we can set upper limits on the mass of the graviton ($m_g \leq 7.7 \times 10^{-23}$ eV/c^2) and on the violation of the Lorentz invariance. Note that, strictly speaking, this is not a complete proof of GR, but a strong statement that *GR is sufficient to describe current GW observation.*

4.4.2   DETECTION OF TWO MERGING NEUTRON STARS

The first merger of two neutron stars was detected by LIGO and Virgo detectors on August 17, 2017, it was a very loud GW signal (with SNR $\approx$ 32). The fact that we had three operating detectors (despite the lower sensitivity of Virgo) allowed a good sky localization of the source and to identify it with the host galaxy NGC 4933 [1]. The GW event GW170817 was also observed in multiple bands by electromagnetic telescopes [8]. Those multi-messenger observations lead to a set of very important discoveries: (i) observation of a gamma-ray burst GRB 170817A by Fermi confirms that the some gamma-ray burst are associated with the merging neutron stars; (ii) simultaneous observations allowed measurement of the luminosity distance (GW) and the redshift (e/m) and independent estimation of the Hubble constant, $H = 70^{+12}_{-8}$ km s^{-1}Mpc^{-1} [7]; and (iii) the time delay between arrival of GW signal and e/m signal was short enough to close a large class of Horndeski-type models of gravity (alternative to GR) [92]. The event was seen from gamma rays to radio with a time-varying delay in a maximum of the emission spectrum after the merger which allowed to probe various theories describing this spectacular event.

The GW observations allowed to estimate the chirp mass with an amazing accuracy $M_c = 1.188^{+0.04}_{-0.001} M_\odot$, however, because we observed only the inspiral part, the measure of the individual masses was rather poor $m_{1,2} \in [0.86, 2.26]$ which is within the expected mass range for the neutron stars. The distance to the system was just ≈ 40 Mpc. The deformability of the neutron stars enters (formally) the GW phase at 5PN order but with a (potentially) large pre-factor and could produce a measurable secular shift in the phase (as compared to binary BH signal), but the effect becomes stronger at higher frequencies (around merger) where current detectors do not have enough sensitivity. The inferred results prefer the "soft" neutron stars equation of state (small tidal deformability) and do not exclude zero deformability (black holes) [1].

It was a luck to have such a strong first signal from merging neutron stars, but the GW signal was contaminated by a very strong instrumental glitch in the Livingston LIGO data, the glitch was a very short in time ≈ 0.1 sec and happened right in the middle of the GW signal. Two approaches were taken to mitigate the glitch in the measurement of GW properties: (i) to window the data and the template around the glitch time (basically cut out a smooth gap) and (ii) to model the glitch using a wavelet reconstruction [42]. In both cases, we have lost a small part of the GW signal, in the first case explicitly, in the second method the signal is removed together with the glitch leaving the noise behind. The right approach would be to extend the model to include glitch and GW signal.

4.4.3   OTHER GW SOURCES IN LIGO-VIRGO BAND

In the previous subsection, we have described four detected coalescing binary BHs. We could also expect a binary system consisting of BH-neutron star. The GW signal produced by those systems is expected to be weaker due to a larger mass ratio. Those systems will be detected with improved sensitivity of the ground-based detectors and during the long observing runs.

Other localized sources of GW waves are the fast-spinning neutron stars with non-zero ellipticity. The perfectly spherical NS will not emit GWs, we need a small "mountain" on the NS's crust to create the non-zero second order derivative in the quadrupole moment.

Current observations set upper limits on ellipticity of neutron stars: $(I_{xx} - I_{zz})/I_{zz}$, see [9] for upper limit for known pulsars. The most optimistic (but not realistic) limit can be obtained by assuming that the rotational energy is released in GWs and responsible for the spin-down rate. The GW upper limit for some pulsar (from O1) is now significantly lower than that optimistic limit. There are several suggestions for the origin of a small "mountain": (i) young non-relaxed neutron star; (ii) processes in the interior coupled to the crust; and (iii) accretion responsible for creating a "hot spot." The GWs from such sources are continuous and, to a high degree, monochromatic (unless there are glitches), their detection requires long integration time. Because the signal is present in the data all the time we need to take into account amplitude and phase modulation caused by the relative motion of the source and the detector. While the targeted GW searches (for known pulsars) could be done on the large scale clusters, the blind, all-sky search requires a lot of computational resources. The special distributed search technique called "Einstein@Home" was set up, where anyone can take part in the search for GWs by delegating the idle time of personal computers (https://einsteinathome.org/).

Example of the non-localized GW source: relic stochastic GW background generated in the early Universe. Those GW are the result of quantum fluctuations of vacuum which were efficiently amplified during the inflationary period (pumping the energy of the expanding Universe into the graviton production). The GW signal is the noise like and therefore could be characterized by power spectral density or, better, as the energy density in GWs. The signal spans across whole frequency band (from nano-Hz to kilo-Hz). There are other processes in the early Universe which could produce potentially detectable stochastic GW background, we will briefly mention them in the next section within the context of LISA sources.

Another important search conducted on the LIGO and VIRGO data is a search for *un-modeled* transient GW signal. The sources of such transient signals are non-spherical supernovae explosion, bursts generated by kinks or cusps formed on the cosmic strings, bursts of radiation from the periapsis passage of two compact objects on a very eccentric (or even parabolic) orbit. In order to assess the significance of those transient signals, it is essential to understand the noise of GW detectors and keep all (or almost all) noise artefacts under control. Alternatively, we utilize the data from two (or more if available) GW detectors to see if the signals are registered in each detector within the light travel time and coherent (consistent amplitude, based on the antenna response function). The first GW event was first detected by the burst search, which is natural given its strength and the short duration.

4.5   GRAVITATIONAL SOURCES IN LISA BAND

In this section we consider the major GW sources in the mHz band. Few examples of those sources are given in Figure 4.10 (taken from [20]).

4.5.1   MASSIVE BLACK HOLE BINARIES

We have already mentioned the massive BH binaries in previous sections. The general belief is that MBHs are hosted in the nuclei of (almost) all galaxies, this is supported by observing the stellar dynamics in the nearby galaxies. The best example is Milky Way where we see the bright O-B class stars on the Keplerian orbits around a dark compact object with estimated mass 4×10^6 $M_\odot$ [69, 71]. The best candidate for this source is an MBH.

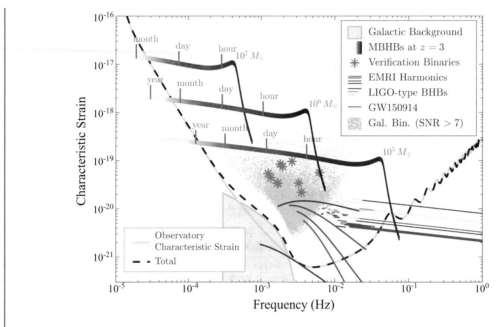

Figure 4.10: Characteristic GW strain of major LISA sources in the frequency domain and the noise rms. Color-coding shows accumulation of SNR over the observation time. The gray region is the stochastic signal created by unresolvable white dwarf binaries.

The initial seeds for the MBHs could have been the remnants from the very first stellar population (pop-III stars); those are "light seed," or could have been produced by a direct collapse of a giant hydrogen cloud, this scenario is referred to as "heavy seeds." In the former case, the seed BHs could be of a mass $\lesssim 10^3 \, M_\odot$, in the latter case the initial BH could be more massive $\sim 10^5 \, M_\odot$. The initial seed BHs have accumulated the mass through (almost constant) gas accretion and the galactic mergers [29, 57, 106]. We do observe merging galaxies, but the problem is to bring two MBHs together close enough so that the gravitational radiation governs the binary evolution. This is possible through the interaction of the binary system with the surrounding stars (the tri-axial potential helps to bring stars close to the MBH binary) and/or through the interaction with the gas, both processes remove the energy from the binary MBH system bringing BHs closer together [64, 107, 145]. As a result of the interaction with the gas and stars the binary could have a non-negligible eccentricity as the binary enters the LISA frequency band. The gas accretion is responsible for the accumulation of the largest part of the MBH mass, and the accreting gas is most likely to form a circum-nuclear disk donating the angular momentum to the MBH in a coherent way. This should lead to a rather high spin of MBHs [29, 118]. Those are two distinct features (non-negligible eccentricity and high spins) we could expect for merging MBH binaries in the LISA's data.

The GW signal from MBHs is very similar to the observed for the solar mass BBHs in the LIGO band. The waveform could be scaled to be a total mass invariant if the overall amplitude M/D_L is factored out (M is a total mass and D_L is a luminosity distance) and time is expressed in units of $M: t \rightarrow t/M$. Note that we have already exploited this in plotting GW signals in Figures 2.4 and 2.5. The GW signal from MBHs is larger in amplitude and longer in duration, the corresponding scaling in the frequency domain can be done by using a dimensionless frequency $f \rightarrow fM$. The rather large amplitude of GW signal from merging MBHs allows us to detect those systems to a very high redshift, in fact, the merging systems with the total mass (in the source frame) between 10^4 $M_\odot$ and 5×10^6 $M_\odot$ will be seen by LISA throughout the Universe. The expected high SNR of those events brings very stringent requirements on the accuracy of GW models for those sources, which is well beyond what is currently used in LIGO/Virgo data analysis. The expected event rate is between few and few hundreds per year [90]. The range reflects the current astrophysical uncertainties about initial MBH seeds, their mass accumulation, and the time delay between the merger of galaxies and the merger of MBHs. Measuring the masses and spins (together with the events rate) can reveal the channel of MBH formation.

We can localize some of those merging binaries to within 10 square degrees on the sky which will allow electromagnetic telescopes to do simultaneous and/or follow-up observations. We do not expect any electromagnetic signal associated with merging BHs ("dry merger") but we could have some non-negligible amount of surrounding gas interacting with the binary systems ("wet merger") which can produce potentially observable electromagnetic counterpart. The measurement of the luminosity distance with GWs from MBHs is usually limited by the weak lensing uncertainties. The GW signal does not provide information about the redshift to the source, however, we could measure the redshift if there is the associated detectable electromagnetic counterpart. The combination of the redshift and the luminosity distance allows us to constrain cosmological parameters such as Hubble constant, dark energy equation of state. In the absence of the strong electromagnetic counterpart, we still can constrain cosmological parameters by applying statistical methods to the population of detected MBH binaries. The statistical method considers each galaxy in the error box (defined by uncertainties in measurements of the sky position and distance) as a potential host of the merger and we cross-correlated the candidates across all detected GW signals imposing the consistency of the underlying cosmological model [104, 127].

The characteristic strain (see discussion around Eq. (4.1)) of GW signal from three MBH binaries is given in Figure 4.10, those are binaries at the distance (redshift) $z = 3$ and with the total mass $10^5, 10^6, 10^7$ $M_\odot$. One can see that the duration of the signal in the band varies between a few weeks and few months.

4.5.2   EXTREME MASS RATIO INSPIRALS

We have already spent some time on the extreme mass ratio inspiral (EMRIs) in Section 2.4.2. Here we will look at it from the astrophysical point of view. To have an EMRI we need a massive BH of mass $10^4 - 10^7$ M$_\odot$ which (as we have discussed above) resides in the galactic nuclei. We have also discussed the formation of MBHs in the previous subsection. However, in this subsection we consider MBHs in ordinary quiescent (relaxed) galaxies and describe the interaction of the central MBH with the stellar environment.

The MBHs in the galactic nuclei should be surrounded by a cluster of stars and/or stellar remnants. When the stars interact gravitationally, they tend to divide the kinetic energy equally and, while equipartition is not entirely reached in practice, this process causes more massive objects to sink deeper in the potential well of the MBH. This process is called mass segregation [31]. As a result, we expect stellar mass BHs to form a steep power-law density cusp around the MBH $n(r) \sim r^\alpha$ with $\alpha \simeq 1.7 - 2.0$, which dominates for $r < 0.1$pc. Those "cuspy" galactic nuclei are dense enough for the efficient 2-body relaxation, i.e., mutual gravitational deflection and contact collisions. As a result of 2-body relaxation, a compact stellar mass object can be deflected onto a very eccentric orbit with a small periapsis distance. During the periapsis passage the system loses energy and angular momentum in bursts of gravitational radiation that makes the system stronger bound. The orbit shrinks and finally the stellar remnant decouples from the cusp and becomes EMRI. The crucial point for this picture is to have compact stellar objects in the "loss cone" (orbits with impact parameter sufficiently small that they can be captured or tidally disrupted by the MBH). Several mechanisms could keep the "loss cone full": triaxiality of the potential due to non-spherical galactic nuclei, presence of a massive perturber (such as a BH with mass $10^2 - 10^4$ M$_\odot$ or a large molecular cloud) in the vicinity of the cusp.

EMRIs are one of the prime sources for LISA observatory. The expected observed rate of EMRIs depends on several ingredients: (i) population of MBHs in the LISA band ($M_\bullet \in [10^5, 10^7]$ M$_\odot$) and their spins; (ii) fraction of MBHs surrounded by a dense stellar cusp; and (iii) the EMRI rate per individual MBH.

The mass function of the MBH formed from the light seeds (see Section 4.5.1) is approximated as [121, 122]:

$$\frac{dn}{d\log M} = 0.005 \left(\frac{M_\bullet}{3 \times 10^6 \, \text{M}_\odot} \right)^{-0.3} \text{Mpc}^{-3}. \tag{4.36}$$

So far, we do not have reliable information about the distribution of the spin of MBHs. X-ray observations of some active galactic nuclei provide information about the spin of accreting MBHs in the center, but those BHs are of higher mass $> 10^7$ M$_\odot$ and embedded in the gaseous circum-nuclear disk. Besides, all present estimations of the spin are heavily model-dependent and could vary significantly depending on the underlying assumptions [35]. The self-consistent model of MBH formation [15, 16, 29, 118] has significant accretion which spins up the MBHs and it predicts that the most MBHs in the LISA range have (dimensionless) spins $a > 0.9$.

As we have mentioned previously, we expect that the MBHs are immersed in a stellar cusp, which is the steady-state solution for the distribution of stars in the gravitational potential of an MBH. However, as discussed in the previous section, the galaxy merge and the cusp is depleted, it is used by a pair of MBHs to reduce the orbital energy and become tightly bound so that the binary evolution is driven mainly by gravitational radiation. It takes some time to re-grow the cusp after the merger of MBHs. This time depends on the parameters of the MBH binary, namely on its total mass M_b and the mass ratio $q = m_1/m_2 > 1$ [22]:

$$t_{\rm cusp} \approx 6 \left(\frac{M_b}{10^6 \, M_\odot} \right)^{1.19} q^{-0.35} \, {\rm Gyr}. \qquad (4.37)$$

To obtain this expression we have used the observed relation between the mass of MBH and the stellar dispersion velocity (so called $M - \sigma$ relation) from [76]. Typical cusp regrowth timescale is a significant fraction of the Hubble time for equal mass binaries with total mass $10^6 \, M_\odot$, whereas they tend to become unimportant for lower mass MBHs (generally less than 1 Gyr for a $10^5 \, M_\odot$ MBH). If the cusp regrowth time is too long, the MBH might undergo another merger which makes it problematic to have EMRIs.

The current estimate for the number of EMRIs per galaxy is given in [14], which accounts for the effect of the mass segregation:

$$R_0 = 300 \left(\frac{M_\bullet}{10^6 \, M_\odot} \right)^{-0.19} \, {\rm Gyr}^{-1}, \qquad (4.38)$$

where $M_\bullet$ is mass of MBH. Note that this is a typical EMRI rate for a Milky Way like galaxy, and care must be taken while extrapolating it to low mass MBHs [22]. In particular, this rate assumes the steady-state supply of stellar remnants relying on that there is a sufficient number in the vicinity to keep the loss cone full.

Within these assumptions, we anticipate observing a few hundred of EMRIs per year with the total SNR above 20. However, this result has a rather large "error bar" coming from the astrophysical uncertainties which lead to the range from few to few thousands GW signals from EMRIs per year.

We can detect EMRIs up to redshift 2–5, depending on the spin of MBH, orbital inclination and the mass of an in-falling compact object (which as we now know from LIGO observations could be as high as 30 $M_\odot$). The sky localization of EMRIs comes primarily from the antenna response function and the Doppler modulation of the phase. The signals are long-lived, which allows localizing the sources on the sky to a few degrees. Detection of 10–20 EMRIs could be used to constrain the Hubble parameter to about 1% using statistical methods described in [96].

EMRIs spend 10^4–10^6 orbits in the close vicinity of an MBH and could be used to infer the physics of MBH, in particular, to test if the central massive object is indeed a Kerr black hole as described by General Relativity theory. There could be several reasons which cause the deviation from the Kerr solution. (i) The EMRI could be coupled to the environment. For example, it could have a gaseous disk surrounding the MBH, there could be a secondary BH on the highly eccentric orbit or a large molecular cloud. (ii) The central object is not a BH but still satisfies GR; the current alternatives to BH are the objects consisting of some exotic matter:

massive boson stars [88] or gravastars [98]. Those objects are very compact but do not have an event horizon. (iii) Finally, the deviation could be due to violation of the General Relativity theory.

We use a compact object orbiting the central massive object as a gravity probe. Similar to the geodesy where the satellite maps the gravitational potential of the Earth, we want to extract the multipolar structure of a central body. The GWs emitted by EMRIs encode a map of the spacetime structure and by observing these gravitational waves with LISA we can precisely characterize the multipole structure. If the central object is a Kerr black hole, then all multipole moments are determined by its mass and spin ("no-hair" theorem):

$$M_l + i S_l = (i a)^l M_\bullet^{l+1}, \tag{4.39}$$

where M_l and S_l are mass and current multipoles, and a, $M_\bullet$ are spin and mass of MBH. If we can measure the first three moments we can, therefore, check whether the central object is consistent with being a Kerr BH. We can determine the deviation in the quadrupole moment from the Kerr value down to $10^{-4} - 10^{-3}$ [22]. We refer the reader to several reviews on the mapping spacetime with EMRIs for the further reading [24, 67].

The strongest harmonics of the GW signal from an EMRI are shown in Figure 4.10.

4.5.3   GALACTIC WHITE DWARF BINARIES

Our Galaxy, the Milky Way, is very bright in the LISA band. We expect 10^6–10^8 Galactic white dwarf binaries emitting GW during the whole mission duration. There are two type of white dwarf binaries: detached binaries which could be seen as two point-masses evolving under gravitational radiation reaction; and binaries with the mass transfer [101]. In the latter case, the orbital evolution depends on the masses of a donor and acceptor and the intensity of the gravitational radiation, so the orbital period could increase or decrease. The vast majority of those binaries emit GWs below few mHz and their frequency evolution stays within a Fourier bin during the entire mission, so they are seen as monochromatic sources. The superposition of all (almost) monochromatic GWs below few mHz will form a stochastic GW signal. Note that this stochastic signal is not isotropic as the majority of white dwarf binaries are confined to the Galactic bulge and plane. The GW stochastic signal is usually above the instrumental noise, however, it has a periodic modulation (with a half year period) due to LISA's orbital motion [59, 137]. Some binaries are sufficiently bright to stand above the stochastic foreground and could be detected individually. Above a few mHz, the number of binaries drops significantly and we start resolving individual signals [34, 45, 94]. Those resolvable GW signals from white dwarf binaries could be subtracted from the data facilitating detection of other (weak) GW signals. It is expected that we should be able to detect and characterize about 15,000 individual sources. About a dozen white dwarf binaries are known from the electromagnetic observations; they are called *verification* binaries. We know their position on the sky and their orbital frequencies; those are *guaranteed* GW sources in the LISA band and they will be used to calibrate the performance of the instrument. Zwicky Transient Facility (ZTF) is performing a survey to identify more such binaries, especially at high frequencies. GAIA and LSST will discover at least a few hundred more white dwarf binaries in the LISA band. Detection of the anisotropic stochastic foreground GW signal and individually resolved systems will allow us to trace and study the overall distribution and formation history of stars in our Milky Way, in its neighboring galaxies and other structures, like globular clusters and the stellar stream.

The stochastic foreground is depicted by a grey shaded area in Figure 4.10, and blue stars are the verification binaries.

4.5.4   STELLAR MASS BINARY BHS

The GW150904 source could have been observed by a LISA-like observatory about 10 years before it enters the LIGO band. The SNR of such a binary is about 6–8 if the initial orbital period is about 10^{-2} Hz and it is being observed over 5 years [120]. LISA will be able to observe a population of stellar-mass black hole binaries within the redshift $z < 0.3$. Many of those binaries are too weak to be detected individually and they will form a stochastic signal. The binaries below 10 mHz and with low chirp mass will not evolve significantly (but not negligibly!) over the LISA mission time. The stochastic GW signal is expected to be largely isotropic with possibly some "hot spots" from the cluster of sources. Some of those binaries will have sufficient SNR to be detected individually and characterized. Those sources spend years in the LISA band and fall in the high-frequency end of its sensitivity (see Figure 4.10). Similar to EMRIs and Galactic binaries, the sky localization will be determined from the Doppler modulation of the signal below a square degree accuracy. Since we observe the early inspiral, we should be able to measure the chirp mass (from the frequency evolution of the signal) with a very high accuracy but will have a weak constraint on the mass ratio. Whether eccentricity is low or not depends on the channel of their formation. We expect a rather low eccentricity if both BHs were born in the binary system (as a result of the stellar evolution). Measuring of the chirp mass allows us to predict the time when the GW signal enters the LIGO (or whatever ground-based GW detector will be operational at that time) band and perform the multi-band observations. We expect some GW signals from the stellar mass binary BHs to sweep across the high end of the LISA band and re-appear after some time within the reach of the ground-based interferometers. The sky location and estimate of the merger time of these binaries should help to identify the corresponding sources in two frequency bands. The cross band observations will significantly increase accuracy in the estimation of the parameters, but most importantly, it will also allow us to test some predictions of the General Relativity theory with extremely high accuracy. In particular, we should be able to probe generation of gravitational radiation (whether there are other channels of energy dissipation), and dispersion of GW signal, as well as, to test the consistency of the inspiral and merger parts of the signal [10, 30].

These binaries are shown as grey and blue (for GW150904-like system) lines at the high-frequency end in Figure 4.10.

4.5.5   OTHER SOURCES IN THE LISA BAND

Another important source in the LISA band is the stochastic GW signal from the early Universe. We have already mentioned it as a LIGO source. The spectrum of this stochastic signal spans across all frequencies from kHz to nHz. There are different mechanisms which could lead to the production of a cosmological gravitational wave background. The primary mechanism is the adiabatic amplification of quantum fluctuations [74], which is most efficient during inflation. The other mechanisms involve classical (as opposed to quantum) sources and have the potential to be much stronger in certain frequency bands. There is a vast literature reviewing this topic [40, 81, 82]. A phase transition corresponding to symmetry breaking of the fundamental interactions could have happened in the early Universe. In the first-order phase transition, the Universe is initially trapped in a metastable phase (with unbroken symmetries). The transition from the metastable phase to the ground state occurs by the quantum tunnelling of a scalar field across a potential energy barrier. This transition nucleates randomly in bubbles. The size of these bubbles increases as the temperature of the Universe drops and large bubbles then collide bringing the Universe to a broken symmetry phase. Gravitational radiation is produced as soon as the spherical symmetry of an individual bubble is broken during the collisions. The spectrum of GWs from the first-order phase transition is peaked around a frequency determined by the typical temperature at which the transition takes place. The electroweak phase transition happened at an energy scale of 100 GeV, for which the peak frequency is ~ 0.1 mHz, in LISA's sensitive frequency band.

Another source of stochastic GW signal is the decay of cosmic string loops [109, 144]. When two long strings reconnect, two long kinked strings are produced. These kinks propagate along the strings and tend to straighten and diminish in strength over time as energy is emitted in gravitational wave radiation. Kinks produce burst-like GWs. A second mechanism for the production of GW bursts from strings are cusps, which are created when a tiny part of the string propagates with a speed close to the speed of light. It was shown in [56] that GW bursts generated by cusps are more likely to be detected in GW observations than those generated by kinks. To detect the GW burst we need to clearly understand behavior of the instrument so we can separate them from the noise artifacts (glitches) and the analysis of the LISA Pathfinder data could help us in doing so [17]. Knowing the shape of the GW burst in time and/or in frequency domain should help us (see Section 3.5) to discriminate the signal of the astrophysical origin from the instrumental/environmental glitch. Alternatively, one can also use the so-called *null-stream data*—the combinations of measurements which are insensitive to the GW signal; in the case of LISA, that would be a Sagnac combination (4.22).

4.6   GRAVITATIONAL SOURCES IN PTA BAND

The GW sources in the PTA band were already described earlier in this book. The prime source is a population of the super-massive BH binaries in the broad orbits. The total mass of those binaries is in the range 10^7–10^9 $M_\odot$ with the orbital period of order a few years. PTA is sensitive only to the sources in the local Universe $z < 1$. We expect a rather numerous population of sources in the nHz band, many of them will not be individually resolved and they will form a stochastic GW signal (similar to one produced by Galactic white dwarf binaries and stellar mass binaries in the LISA band) (see [119] and references therein). Like in the case of Galactic binaries in the LISA band, some of the nearby sources could stand above the stochastic signal and could be detected individually. The MBH binaries at the low-frequency end of the sensitivity band do not evolve (chirp) appreciably within the observation time and produce continuous monochromatic GWs, the sources at high frequency (10^{-7} Hz and above) could have a measurable drift in frequency. Because these MBH binaries are on the broad orbits they might have non-negligible interaction with the stellar environment, thus the orbital dynamics could be altered from just gravitational wave-driven evolution. If coupling to the environment takes place, then the expected spectrum of the stochastic GW signal will be somewhat suppressed at low frequencies (where PTA is most sensitive). A similar effect could be observed if binaries are on the eccentric orbits: the maximum of the spectrum of gravitational radiation from eccentric binaries is shifted to higher frequencies (higher than two harmonics of the orbital frequency). Both eccentricity and interaction with the stellar environment make it harder to detect the GW signal from the population of MBH binaries. The International Pulsar Timing Array consortium makes rapid progress in assembling the data and searching for both components (stochastic signal from the population and individual sources) [113, 143].

Another stochastic GW signal which spans down to the PTA band was already described; this is the signal produced in the early Universe and/or by decaying cosmic string loops.

While there is no yet detected GW signals by PTA, the upper limits set by the current observations start to rule out some optimistic astrophysical models [123, 132]. We refer the reader to a very nice review article on detecting stochastic GW signal by PTA in [112].

The GW signals in the PTA band are long-lived and weak; they will slowly emerge from the noise as we accumulate more pulsars and longer observational span. It is therefore important to assess the statistical significance of the first (potential) detection with PTA. We cannot perform the time slides (described in Section 3.5.3); instead, we modify the search parameters so that they are insensitive to the GW signal. This method is described in [44, 130] and we will not go into any details here.

We also could expect a burst-like GW event in the PTA band which comes from the *memory effect* [41, 63]. The nonlinear memory effect is a slowly-growing, non-oscillatory contribution to the gravitational-wave amplitude. It originates from the gravitational waves that are sourced by the previously emitted waves. This effect could be seen as a coherent jump in the observed residuals in the PTA data [140].

Bibliography

[1] B. P. Abbott, R. Abbott, T. D. Abbott, F. Acernese, K. Ackley, C. Adams, T. Adams, P. Addesso, R. X. Adhikari, V. B. Adya, et al. GW170817: Observation of gravitational waves from a binary neutron star inspiral. *Physical Review Letters*, 119(16):161101, October 2017. DOI: 10.1103/PhysRevLett.119.161101 81, 82

[2] B. P. Abbott et al. Binary black hole mergers in the first advanced LIGO observing run. 2016. DOI: 10.1103/PhysRevX.6.041015 78

[3] B. P. Abbott et al. GW151226: Observation of gravitational waves from a 22-solar-mass binary black hole coalescence. *Physical Review Letters*, 116(24):241103, 2016. 4, 79

[4] B. P. Abbott et al. Observation of gravitational waves from a binary black hole merger. *Physical Review Letters*, 116(6):061102, 2016. DOI: 10.1142/9789814699662_0011 4, 77

[5] B. P. Abbott et al. Properties of the binary black hole merger GW150914. *Physical Review Letters*, 116(24):241102, 2016. DOI: 10.1103/PhysRevLett.116.241102

[6] B. P. Abbott et al. Tests of general relativity with GW150914. *Physical Review Letters*, 116(22):221101, 2016. 77, 81

[7] B. P. Abbott et al. A gravitational-wave standard siren measurement of the Hubble constant. *Nature*, 551(7678):85–88, 2017. DOI: 10.1038/nature24471 82

[8] B. P. Abbott et al. Multi-messenger observations of a binary neutron star merger. *Astrophysical Journal*, 848(2):L12, 2017. DOI: 10.3847/2041-8213/aa91c9 81

[9] B. P. Abbott et al. First search for gravitational waves from known pulsars with advanced LIGO. *Astrophysical Journal*, 839(1):12, 2017. DOI: 10.3847/1538-4357/aa677f [Erratum: *Astrophysical Journal*, 851(1):71, (2017)]. 82

[10] B. P. Abbott et al. GW170104: Observation of a 50-solar-mass binary black hole coalescence at redshift 0.2. *Physical Review Letters*, 118(22):221101, 2017. 78, 79, 81, 89

[11] P. A. R. Ade et al. A measurement of the cosmic microwave background B-mode polarization power spectrum at sub-degree scales with POLARBEAR. *Astrophysical Journal*, 794(2):171, 2014. DOI: 10.1088/0004-637x/794/2/171 5

[12] P. Ajith et al. A Template bank for gravitational waveforms from coalescing binary black holes. I. Non-spinning binaries. *Physical Review*, D77:104017, 2008. DOI: 10.1103/physrevd.77.104017 [Erratum: *Physical Review*, D79:129901, (2009)]. 35

[13] B. Allen. A χ^2 time-frequency discriminator for gravitational wave detection. *Physical Review*, D71:062001, 2005. DOI: 10.1103/physrevd.71.062001 57, 60

[14] P. Amaro-Seoane and M. Preto. The impact of realistic models of mass segregation on the event rate of extreme-mass ratio inspirals and cusp re-growth. *Classical and Quantum Gravity*, 28(9):094017, May 2011. DOI: 10.1088/0264-9381/28/9/094017 87

[15] F. Antonini, E. Barausse, and J. Silk. The coeution of nuclear star clusters, massive black holes, and their host galaxies. *Astrophysical Journal*, 812:72, October 2015. DOI: 10.1088/0004-637x/812/1/72 86

[16] F. Antonini, E. Barausse, and J. Silk. The imprint of massive black hole mergers on the correlation between nuclear star clusters and their host galaxies. *Astrophysical Journal Letters*, 806:L8, June 2015. DOI: 10.1088/2041-8205/806/1/l8 86

[17] M. Armano, H. Audley, G. Auger, J. T. Baird, M. Bassan, P. Binetruy, M. Born, D. Bortoluzzi, N. Brandt, M. Caleno, L. Carbone, A. Cavalleri, A. Cesarini, G. Ciani, G. Congedo, A. M. Cruise, K. Danzmann, M. de Deus Silva, R. De Rosa, M. Diaz-Aguiló, L. Di Fiore, I. Diepholz, G. Dixon, R. Dolesi, N. Dunbar, L. Ferraioli, V. Ferroni, W. Fichter, E. D. Fitzsimons, R. Flatscher, M. Freschi, A. F. García Marín, C. García Marirrodriga, R. Gerndt, L. Gesa, F. Gibert, D. Giardini, R. Giusteri, F. Guzmán, A. Grado, C. Grimani, A. Grynagier, J. Grzymisch, I. Harrison, G. Heinzel, M. Hewitson, D. Hollington, D. Hoyland, M. Hueller, H. Inchauspé, O. Jennrich, P. Jetzer, U. Johann, B. Johlander, N. Karnesis, B. Kaune, N. Korsakova, C. J. Killow, J. A. Lobo, I. Lloro, L. Liu, J. P. López-Zaragoza, R. Maarschalkerweerd, D. Mance, V. Martín, L. Martin-Polo, J. Martino, F. Martin-Porqueras, S. Madden, I. Mateos, P. W. McNamara, J. Mendes, L. Mendes, A. Monsky, D. Nicolodi, M. Nofrarias, S. Paczkowski, M. Perreur-Lloyd, A. Petiteau, P. Pivato, E. Plagnol, P. Prat, U. Ragnit, B. Raïs, J. Ramos-Castro, J. Reiche, D. I. Robertson, H. Rozemeijer, F. Rivas, G. Russano, J. Sanjuán, P. Sarra, A. Schleicher, D. Shaul, J. Slutsky, C. F. Sopuerta, R. Stanga, F. Steier, T. Sumner, D. Texier, J. I. Thorpe, C. Trenkel, M. Tröbs, H. B. Tu, D. Vetrugno, S. Vitale, V. Wand, G. Wanner, H. Ward, C. Warren, P. J. Wass, D. Wealthy, W. J. Weber, L. Wissel, A. Wittchen, A. Zambotti, C. Zanoni, T. Ziegler, and P. Zweifel. Sub-Femto-*g* free fall for space-based gravitational wave observatories: Lisa pathfinder results. *Physical Review Letters*, 116:231101, June 2016. DOI: 10.1103/physrevlett.116.231101 5, 90

[18] R. L. Arnowitt, S. Deser, and C. W. Misner. The dynamics of general relativity. *General Relativity Gravitation*, 40:1997–2027, 2008. 24

[19] K. G. Arun, A. Buonanno, G. Faye, and E. Ochsner. Higher-order spin effects in the amplitude and phase of gravitational waveforms emitted by inspiraling compact binaries: Ready-to-use gravitational waveforms. *Physical Review*, D79:104023, 2009. DOI: 10.1103/physrevd.79.104023 [Erratum: *Physical Review*, D84:049901, (2011)]. 26

[20] H. Audley et al. Laser interferometer space antenna. arXiv:1702.00786, 2017. 83

[21] S. Babak.   Building a stochastic template bank for detecting massive black hole binaries. *Classical and Quantum Gravity*, 25:195011, 2008. DOI: 10.1088/0264-9381/25/19/195011 46

[22] S. Babak, J. Gair, A. Sesana, E. Barausse, C. F. Sopuerta, C. P. L. Berry, E. Berti, P. Amaro-Seoane, A. Petiteau, and A. Klein. Science with the space-based interferometer LISA. V: Extreme mass-ratio inspirals. *Physical Review*, D95(10):103012, 2017. DOI: 10.1103/physrevd.95.103012 87, 88

[23] S. Babak, J. R. Gair, and R. H. Cole. Extreme mass ratio inspirals: Perspectives for their detection. *Fundamental Theories of Physics*, 179:783–812, 2015. DOI: 10.1007/978-3-319-18335-0_23 29

[24] S. Babak, J. R. Gair, A. Petiteau, and A. Sesana. Fundamental physics and cosmology with LISA. *Classical and Quantum Gravity*, 28:114001, 2011. DOI: 10.1088/0264-9381/28/11/114001 88

[25] S. Babak and A. Sesana. Resolving multiple supermassive black hole binaries with pulsar timing arrays. *Physical Review*, D85:044034, 2012. DOI: 10.1103/physrevd.85.044034 44, 76

[26] S. Babak, A. Taracchini, and A. Buonanno. Validating the effective-one-body model of spinning, precessing binary black holes against numerical relativity. *Physical Review*, D95(2):024010, 2017. DOI: 10.1103/physrevd.95.024010 35

[27] S. Balmelli and T. Damour. New effective-one-body Hamiltonian with next-to-leading order spin-spin coupling. *Physical Review*, D92(12):124022, 2015. DOI: 10.1103/physrevd.92.124022 32

[28] L. Barack and A. Pound. Self-force and radiation reaction in general relativity. *Reports on Progress in Physics*, 82:016904, 2019. DOI: 10.1088/1361-6633/aae552 29

[29] E. Barausse.   The eution of massive black holes and their spins in their galactic hosts. *Monthly Notices of the Royal Astronomical Society*, 423:2533–2557, 2012. DOI: 10.1111/j.1365-2966.2012.21057.x 84, 86

[30] E. Barausse, N. Yunes, and K. Chamberlain. Theory-agnostic constraints on black-hole dipole radiation with multiband gravitational-wave astrophysics. *Physical Review Letters*, 116(24):241104, 2016. DOI: 10.1103/physrevlett.116.241104 89

[31] J. Binney and S. Tremaine. *Galactic Dynamics*. Princeton University Press, 1987. 86

[32] L. Blanchet. Gravitational radiation from post-Newtonian sources and inspiralling compact binaries. *Living Reviews in Relativity*, 17(2), 2014. DOI: 10.12942/lrr-2002-3 16, 23, 24, 25

[33] L. Blanchet, T. Damour, and G. Esposito-Farese. Dimensional regularization of the third post Newtonian dynamics of point particles in harmonic coordinates. *Physical Review*, D69:124007, 2004. DOI: 10.1103/physrevd.69.124007 24

[34] A. Blaut, S. Babak, and A. Krolak. Mock LISA data challenge for the galactic white dwarf binaries. *Physical Review*, D81:063008, 2010. DOI: 10.1103/physrevd.81.063008 88

[35] L. Brenneman. Measuring supermassive black hole spins in active galactic nuclei. arXiv:1309.6334, 2013. DOI: 10.14311/ap.2013.53.0652 86

[36] V. Brumberg. *Essential Relativistic Celestial Mechanics*. CRC Press, 1991. DOI: 10.1201/9780203756591 28

[37] A. Buonanno and T. Damour. Effective one-body approach to general relativistic two-body dynamics. *Physical Review*, D59:084006, 1999. DOI: 10.1103/physrevd.59.084006 29, 30, 31

[38] A. Buonanno, Y.-B. Chen, and M. Vallisneri. Detection template families for gravitational waves from the final stages of binary-black-hole inspirals: Nonspinning case. *Physical Review*, D67:024016, 2003. DOI: 10.1103/physrevd.67.024016 [Erratum: *Physical Review*, D74:029903, (2006)]. 26

[39] A. Buonanno and T. Damour. Transition from inspiral to plunge in binary black hole coalescences. *Physical Review*, D62:064015, 2000. DOI: 10.1103/physrevd.62.064015 29, 32

[40] S. Chongchitnan and G. Efstathiou. Prospects for direct detection of primordial gravitational waves. *Physical Review*, D73:083511, 2006. DOI: 10.1103/physrevd.73.083511 90

[41] D. Christodoulou. Nonlinear nature of gravitation and gravitational wave experiments. *Physical Review Letters*, 67:1486–1489, 1991. DOI: 10.1103/physrevlett.67.1486 91

[42] N. J. Cornish and T. B. Littenberg. BayesWave: Bayesian inference for gravitational wave bursts and instrument glitches. *Classical and Quantum Gravity*, 32(13):135012, 2015. DOI: 10.1088/0264-9381/32/13/135012 82

[43] N. J. Cornish and E. K. Porter. Catching supermassive black hole binaries without a net. *Physical Review*, D75:021301, 2007. DOI: 10.1103/physrevd.75.021301 44

[44] N. J. Cornish and L. Sampson. Towards robust gravitational wave detection with pulsar timing arrays. *Physical Review*, D93(10):104047, 2016. DOI: 10.1103/physrevd.93.104047 91

[45] J. Crowder and N. Cornish. A solution to the galactic foreground problem for LISA. *Physical Review*, D75:043008, 2007. DOI: 10.1103/physrevd.75.043008 88

[46] T. Damour and G. Schaefer. Higher order relativistic periastron advances and binary pulsars. *Nuovo Cimento*, B101:127, 1988. DOI: 10.1007/bf02828697 30

[47] T. Damour. The general relativistic two body problem and the effective one body formalism. *Fundamental Theories of Physics*, 177:111–145, 2014. DOI: 10.1007/978-3-319-06349-2_5 29, 32, 34

[48] T. Damour, A. Gopakumar, and B. R. Iyer. Phasing of gravitational waves from inspiralling eccentric binaries. *Physical Review*, D70:064028, 2004. DOI: 10.1103/physrevd.70.064028 26

[49] T. Damour, B. R. Iyer, and A. Nagar. Improved resummation of post-Newtonian multipolar waveforms from circularized compact binaries. *Physical Review*, D79:064004, 2009. DOI: 10.1103/physrevd.79.064004 33

[50] T. Damour, B. R. Iyer, and B. S. Sathyaprakash. A Comparison of search templates for gravitational waves from binary inspiral. *Physical Review*, D63:044023, 2001. DOI: 10.1103/physrevd.66.027502 [Erratum: *Physical Review*, D72:029902, (2005)]. 26

[51] T. Damour, P. Jaranowski, and G. Schaefer. On the determination of the last stable orbit for circular general relativistic binaries at the third post-Newtonian approximation. *Physical Review*, D62:084011, 2000. DOI: 10.1103/physrevd.62.084011 31

[52] T. Damour, P. Jaranowski, and G. Schaefer. Poincare invariance in the ADM Hamiltonian approach to the general relativistic two-body problem. *Physical Review*, D62:021501, 2000. DOI: 10.1103/physrevd.62.021501 [Erratum: *Physical Review*, D63:029903, (2001)]. 30

[53] T. Damour, P. Jaranowski, and G. Schäfer. Nonlocal-in-time action for the fourth post-Newtonian conservative dynamics of two-body systems. *Physical Review*, D89(6):064058, 2014. DOI: 10.1103/physrevd.89.064058 30

[54] T. Damour and A. Nagar. A new analytic representation of the ringdown waveform of coalescing spinning black hole binaries. *Physical Review*, D90(2):024054, 2014. DOI: 10.1103/physrevd.90.024054 35

[55] T. Damour and A. Nagar. New effective-one-body description of coalescing nonprecessing spinning black-hole binaries. *Physical Review*, D90(4):044018, 2014. DOI: 10.1103/phys-revd.90.044018 32

[56] T. Damour and A. Vilenkin. Gravitational wave bursts from cosmic strings. *Physical Review Letters*, 85:3761–3764, 2000. DOI: 10.1103/physrevlett.85.3761 90

[57] G. De Lucia and J. Blaizot. The hierarchical formation of the brightest cluster galaxies. *Monthly Notices of the Royal Astronomical Society*, 375:2–14, 2007. DOI: 10.1111/j.1365-2966.2006.11287.x 84

[58] P. Diener, I. Vega, B. Wardell, and S. Detweiler. Self-consistent orbital eution of a particle around a Schwarzschild black hole. *Physical Review Letters*, 108:191102, 2012. DOI: 10.1103/physrevlett.108.191102 29

[59] J. A. Edlund, M. Tinto, A. Krolak, and G. Nelemans. The white dwarf—white dwarf galactic background in the LISA data. *Physical Review*, D71:122003, 2005. DOI: 10.1103/physrevd.71.122003 88

[60] A. Einstein. Über gravitationswellen. *Königlich Preussische Akademie der Wissenschaften (Berlin). Sitzungsberichte*, pp. 154–167, 1918. DOI: 10.1002/3527608958.ch12 22

[61] A. Einstein, L. Infeld, and B. Hoffmann. The gravitational equations and the problem of motion. *Annals of Mathematics*, 39:65–100, 1938. DOI: 10.2307/1969015 24

[62] F. B. Estabrook and H. D. Wahlquist. Response of Doppler spacecraft tracking to gravitational radiation. *General Relativity and Gravitation*, 6:439, 1975. DOI: 10.1007/bf00762449 64

[63] M. Favata. The gravitational-wave memory effect. *Classical and Quantum Gravity*, 27:084036, 2010. DOI: 10.1088/0264-9381/27/8/084036 91

[64] D. Fiacconi, L. Mayer, R. Roskar, and M. Colpi. Massive black hole pairs in clumpy, self-gravitating circumnuclear disks: Stochastic orbital decay. *Astrophysical Journal*, 777:L14, 2013. DOI: 10.1088/2041-8205/777/1/l14 84

[65] T. Futamase. The strong field point particle limit and the equations of motion in the binary pulsar. *Physical Review*, D36:321, 1987. DOI: 10.1103/physrevd.36.321 24

[66] J. R. Gair, E. E. Flanagan, S. Drasco, T. Hinderer, and S. Babak. Forced motion near black holes. *Physical Review*, D83:044037, 2011. DOI: 10.1103/physrevd.83.044037 28

[67] J. R. Gair, M. Vallisneri, S. L. Larson, and J. G. Baker. Testing general relativity with low-frequency, space-based gravitational-wave detectors. *Living Reviews in Relativity*, 16:7, 2013. DOI: 10.12942/lrr-2013-7 88

[68] K. Ganz, W. Hikida, H. Nakano, N. Sago, and T. Tanaka. Adiabatic eution of three "constants" of motion for greatly inclined orbits in Kerr spacetime. *Progress of Theoretical Physics*, 117:1041–1066, 2007. DOI: 10.1143/ptp.117.1041 28

[69] A. M. Ghez, S. Salim, N. N. Weinberg, J. R. Lu, T. Do, J. K. Dunn, K. Matthews, M. R. Morris, S. Yelda, E. E. Becklin, T. Kremenek, M. Milosavljevic, and J. Naiman. Measuring distance and properties of the milky way's central supermassive black hole with stellar orbits. *Astrophysical Journal*, 689:1044–1062, December 2008. DOI: 10.1086/592738 63, 83

[70] W. R. Gilks, S. Richardson, and D. Spiegelhalter. *Markov Chain Monte Carlo in Practice*. Chapman & Hall/CRC Interdisciplinary Statistics. Taylor & Francis, 1995. DOI: 10.2307/1271145 51, 55

[71] S. Gillessen, F. Eisenhauer, S. Trippe, T. Alexander, R. Genzel, F. Martins, and T. Ott. Monitoring stellar orbits around the massive black hole in the galactic center. *Astrophysical Journal*, 692:1075–1109, February 2009. DOI: 10.1088/0004-637x/692/2/1075 63, 83

[72] K. Glampedakis. Extreme mass ratio inspirals: LISA's unique probe of black hole gravity. *Classical and Quantum Gravity*, 22:S605–S659, 2005. DOI: 10.1088/0264-9381/22/15/004 27

[73] H. Goldstein. *Classical Mechanics*. Addison-Wesley, 1980. DOI: 10.2307/3610571 28

[74] L. P. Grishchuk. Amplification of gravitational waves in an isotropic universe. *Journal of Experimental and Theoretical Physics*, 40:409, 1975. 7, 90

[75] L. P. Grishchuk and S. M. Kopeikin. The motion of a pair of gravitating bodies including the radiation reaction force. *Soviet Astronomy Letters*, 9:230–232, April 1983. 24

[76] K. Gultekin, D. O. Richstone, K. Gebhardt, T. R. Lauer, S. Tremaine, et al. The M-sigma and M-L relations in galactic bulges and determinations of their intrinsic scatter. *Astrophysical Journal*, 698:198–221, 2009. DOI: 10.1088/0004-637x/698/1/198 87

[77] M. Hannam, P. Schmidt, A. Bohé, L. Haegel, S. Husa, F. Ohme, G. Pratten, and M. Pürrer. Simple model of complete precessing black-hole-binary gravitational waveforms. *Physical Review Letters*, 113(15):151101, 2014. DOI: 10.1103/physrevlett.113.151101 35

[78] I. W. Harry, B. Allen, and B. S. Sathyaprakash. A stochastic template placement algorithm for gravitational wave data analysis. *Physical Review*, D80:104014, 2009. DOI: 10.1103/physrevd.80.104014 46

[79] W. K. Hastings. Monte Carlo sampling methods using Markov chains and their applications. *Biometrika*, 57(1):97–109, 1970. DOI: 10.2307/2334940 53

[80] S. W. Hawking and G. F. R. Ellis. *The Large Scale Structure of Space-Time*. Cambridge University Press, 1973. DOI: 10.1017/cbo9780511524646 3, 4

[81] C. J. Hogan. Gravitational wave sources from new physics. *AIP Conference Proceedings*, 873:30–40, 2006. DOI: 10.1063/1.2405019 90

[82] S. A. Hughes. Listening to the universe with gravitational-wave astronomy. *Annals of Physics*, 303:142–178, 2003. DOI: 10.1016/s0003-4916(02)00025-8 90

[83] R. A. Isaacson. Gravitational radiation in the limit of high frequency. II. Nonlinear terms and the effective stress tensor. *Physical Review*, 166:1272–1279, 1968. DOI: 10.1103/physrev.166.1272 17

[84] Y. Itoh and T. Futamase. New derivation of a third post-Newtonian equation of motion for relativistic compact binaries without ambiguity. *Physical Review*, D68:121501, 2003. DOI: 10.1103/physrevd.68.121501 24

[85] P. Jaranowski, A. Krolak, and B. F. Schutz. Data analysis of gravitational—wave signals from spinning neutron stars. I. The Signal and its detection. *Physical Review*, D58:063001, 1998. DOI: 10.1103/physrevd.58.063001 44

[86] P. Jaranowski and G. Schaefer. Third post-Newtonian higher order ADM Hamilton dynamics for two-body point mass systems. *Physical Review*, D57:7274–7291, 1998. DOI: 10.1103/physrevd.57.7274. [Erratum: *Physical Review*, D63:029902, (2001)]. 24

[87] D. Kennefick and A. Ori. Radiation reaction induced eution of circular orbits of particles around Kerr black holes. *Physical Review*, D53:4319–4326, 1996. DOI: 10.1103/physrevd.53.4319 28

[88] M. Kesden, J. Gair, and M. Kamionkowski. Gravitational-wave signature of an inspiral into a supermassive horizonless object. *Physical Review*, D71:044015, 2005. DOI: 10.1103/physrevd.71.044015 88

[89] S. Khan, S. Husa, M. Hannam, F. Ohme, M. Pürrer, X. J. Forteza, and A. Bohé. Frequency-domain gravitational waves from nonprecessing black-hole binaries. II. A phenomenological model for the advanced detector era. *Physical Review*, D93(4):044007, 2016. DOI: 10.1103/physrevd.93.044007 35

[90] A. Klein et al. Science with the space-based interferometer eLISA: Supermassive black hole binaries. *Physical Review*, D93(2):024003, 2016. DOI: 10.1103/physrevd.93.024003 85

[91] L. D. Landau and E. M. Lifshitz. *The Classical Theory of Fields: Volume 2 (Course of Theoretical Physics)*. Pergamon Press, 1971. 13, 22

[92] D. Langlois, R. Saito, D. Yamauchi, and K. Noui. Scalar-tensor theories and modified gravity in the wake of GW170817. *Physical Review*, D97(6):061501, 2018. DOI: 10.1103/physrevd.97.061501 82

[93] G. P. Lepage. Vegas: An adaptive multidimensional integration program. *CLNS–80/447 (CERN Report Number)*, 1980. 51, 52

[94] T. B. Littenberg. A detection pipeline for galactic binaries in LISA data. *Physical Review*, D84:063009, 2011. DOI: 10.1103/physrevd.84.063009 88

[95] D. R. Lorimer and M. Kramer. *Handbook of Pulsar Astronomy*. December 2004. 62, 72

[96] C. L. MacLeod and C. J. Hogan. Precision of Hubble constant derived using black hole binary absolute distances and statistical redshift information. *Physical Review*, D77:043512, 2008. DOI: 10.1103/physrevd.77.043512 87

[97] G. M. Manca and M. Vallisneri. Cover art: Issues in the metric-guided and metric-less placement of random and stochastic template banks. *Physical Review*, D81:024004, 2010. DOI: 10.1103/physrevd.81.024004 46

[98] P. O. Mazur and E. Mottola. Gravitational condensate stars: An alternative to black holes. arXiv:gr-qc/0109035, 2001. 88

[99] C. Messenger, R. Prix, and M. A. Papa. Random template banks and relaxed lattice coverings. *Physical Review*, D79:104017, 2009. DOI: 10.1103/physrevd.79.104017 46

[100] C. W. Misner, K. S. Thorne, and J. A. Wheeler. *Gravitation*, 2nd ed., W. H. Freeman and Company, 1973. 1, 3, 4, 13, 22

[101] G. Nelemans, S. F. P. Zwart, F. Verbunt, and L. R. Yungelson. Population synthesis for double white dwarfs. II. Semi-detached systems: AM CVn stars. *Astronomy & Astrophysics*, 368:939–949, 2001. DOI: 10.1051/0004-6361:20010049 88

[102] A. Pai, K. Rajesh Nayak, S. V. Dhurandhar, and J. Y. Vinet. Time delay interferometry and LISA optimal sensitivity. In *38th Rencontres de Moriond on Gravitational Waves and Experimental Gravity*, Les Arcs, Savoie, France, March 22–29, 2003. 71

[103] Y. Pan, A. Buonanno, A. Taracchini, L. E. Kidder, A. H. Mroué, H. P. Pfeiffer, M. A. Scheel, and B. Szilágyi. Inspiral-merger-ringdown waveforms of spinning, precessing black-hole binaries in the effective-one-body formalism. *Physical Review*, D89(8):084006, 2014. DOI: 10.1103/physrevd.89.084006 32

[104] A. Petiteau, S. Babak, and A. Sesana. Constraining the dark energy equation of state using LISA observations of spinning massive black hole binaries. *Astrophysical Journal*, 732:82, 2011. DOI: 10.1088/0004-637x/732/2/82 85

[105] A. Petiteau, Y. Shang, S. Babak, and F. Feroz. The search for spinning black hole binaries in mock LISA data using a genetic algorithm. *Physical Review*, D81:104016, 2010. DOI: 10.1103/physrevd.81.104016 50

[106] E. Pezzulli, M. Volonteri, R. Schneider, and R. Valiante. The sustainable growth of the first black holes. *Montly Notices of Royal Astronomical Society*, 471(1):589–595, 2017. DOI: 10.1093/mnras/stx1640 84

[107] H. Pfister, A. Lupi, P. R. Capelo, M. Volonteri, J. M. Bellovary, and M. Dotti. The birth of a supermassive black hole binary. *Montly Notices of Royal Astronomical Society*, 471(3):3646–3656, 2017. DOI: 10.1093/mnras/stx1853 84

[108] E. Poisson, A. Pound, and I. Vega. The motion of point particles in curved spacetime. *Living Reviews in Relativity*, 14(7), 2011. DOI: 10.12942/lrr-2011-7 28, 29

[109] J. Polchinski. Cosmic string loops and gravitational radiation. arXiv:0707.0888, 2007. 90

[110] A. Pound and E. Poisson. Osculating orbits in Schwarzschild spacetime, with an application to extreme mass-ratio inspirals. *Physical Review*, D77:044013, 2008. DOI: 10.1103/physrevd.77.044013 28

[111] T. A. Prince, M. Tinto, S. L. Larson, and J. W. Armstrong. The LISA optimal sensitivity. *Physical Review*, D66:122002, 2002. DOI: 10.1103/physrevd.66.122002 71

[112] J. D. Romano and N. J. Cornish. Detection methods for stochastic gravitational-wave backgrounds: A unified treatment. *Living Reviews in Relativity*, 20:2, 2017. DOI: 10.1007/s41114-017-0004-1 91

[113] P. A. Rosado, A. Sesana, and J. Gair. Expected properties of the first gravitational wave signal detected with pulsar timing arrays. *Monthly Notices of the Royal Astronomical Society*, 451(3):2417–2433, 2015. DOI: 10.1093/mnras/stv1098 91

[114] L. J. Rubbo, N. J. Cornish, and O. Poujade. Forward modeling of space borne gravitational wave detectors. *Physical Review*, D69:082003, 2004. DOI: 10.1103/physrevd.69.082003 64

[115] L. Santamaria et al. Matching post-Newtonian and numerical relativity waveforms: Systematic errors and a new phenomenological model for non-precessing black hole binaries. *Physical Review*, D82:064016, 2010. DOI: 10.1103/physrevd.82.064016 35

[116] B. Schutz. *A First Course in General Relativity*, 2nd ed., Cambridge University Press, 2009. DOI: 10.1017/cbo9780511984181 1, 13

[117] G. Schäfer and P. Jaranowski. Hamiltonian formulation of general relativity and post-Newtonian dynamics of compact binaries. *Living Reviews in Relativity*, 21, 2018. DOI: 10.1007/s41114-018-0016-5 24

[118] A. Sesana, E. Barausse, M. Dotti, and E. M. Rossi. Linking the spin eution of massive black holes to galaxy kinematics. *Astrophysical Journal*, 794:104, 2014. DOI: 10.1088/0004-637x/794/2/104 84, 86

[119] A. Sesana. Pulsar timing arrays and the challenge of massive black hole binary astrophysics. *Astrophysics and Space Science Proceedings*, 40:147–165, 2015. DOI: 10.1007/978-3-319-10488-1_13 91

[120] A. Sesana. Prospects for multiband gravitational-wave astronomy after GW150914. *Physical Review Letters*, 116(23):231102, 2016. DOI: 10.1103/physrevlett.116.231102 89

[121] F. Shankar. Black hole demography: From scaling relations to models. *Classical and Quantum Gravity*, 30(24):244001, December 2013. DOI: 10.1088/0264-9381/30/24/244001 86

[122] F. Shankar, D. H. Weinberg, and J. Miralda-Escudé. Self-consistent models of the AGN and black hole populations: Duty cycles, accretion rates, and the mean radiative efficiency. *Astrophysical Journal*, 690:20–41, January 2009. DOI: 10.1088/0004-637x/690/1/20 86

[123] R. M. Shannon et al. Gravitational waves from binary supermassive black holes missing in pulsar observations. *Science*, 349(6255):1522–1525, 2015. DOI: 10.1126/science.aab1910 91

[124] D. S. Sivia and J. Skilling. *Data Analysis: A Bayesian Tutorial*. Oxford Science Publications, Oxford University Press, 2006. 40

[125] J. Skilling. Nested sampling for general Bayesian computation. *Bayesian Analysis*, 1(4):833–859, December 2006. DOI: 10.1214/06-ba127 51, 55

[126] B. Szilágyi, J. Blackman, A. Buonanno, A. Taracchini, H. P. Pfeiffer, M. A. Scheel, T. Chu, L. E. Kidder, and Y. Pan. Approaching the post-Newtonian regime with numerical relativity: A compact-object binary simulation spanning 350 gravitational-wave cycles. *Physical Review Letters*, 115(3):031102, 2015. DOI: 10.1103/physrevlett.115.031102 34

[127] N. Tamanini, C. Caprini, E. Barausse, A. Sesana, A. Klein, and A. Petiteau. Science with the space-based interferometer eLISA. III: Probing the expansion of the universe using gravitational wave standard sirens. *JCAP*, 1604(04):002, 2016. DOI: 10.1088/1475-7516/2016/04/002 85

[128] A. Taracchini et al. Effective-one-body model for black-hole binaries with generic mass ratios and spins. *Physical Review*, D89(6):061502, 2014. DOI: 10.1103/phys-revd.89.061502 32

[129] A. Taracchini, Y. Pan, A. Buonanno, E. Barausse, M. Boyle, T. Chu, G. Lovelace, H. P. Pfeiffer, and M. A. Scheel. Prototype effective-one-body model for nonprecessing spinning inspiral-merger-ringdown waveforms. *Physical Review*, D86:024011, 2012. DOI: 10.1103/physrevd.86.024011 32, 34

[130] S. R. Taylor, L. Lentati, S. Babak, P. Brem, J. R. Gair, A. Sesana, and A. Vecchio. All correlations must die: Assessing the significance of a stochastic gravitational-wave background in pulsar-timing arrays. *Physical Review*, D95(4):042002, 2017. DOI: 10.1103/phys-revd.95.042002 91

[131] S. R. Taylor, J. R. Gair, and L. Lentati. Using swarm intelligence to accelerate pulsar timing analysis. arXiv:1210.3489, 2012. 51

[132] S. R. Taylor, J. Simon, and L. Sampson. Constraints on the dynamical environments of supermassive black-hole binaries using pulsar-timing arrays. *Physical Review Letters*, 118(18):181102, 2017. DOI: 10.1103/physrevlett.118.181102 91

[133] M. Tessmer and G. Schaefer. Full-analytic frequency-domain gravitational wave forms from eccentric compact binaries to 2PN accuracy. *Annalen der Physik*, 523:813–864, 2011. DOI: 10.1002/andp.201100007 26

[134] S. A. Teukolsky. Perturbations of a rotating black hole. I. Fundamental equations for gravitational electromagnetic and neutrino field perturbations. *Astrophysical Journal*, 185:635–647, 1973. DOI: 10.1086/152444 28

[135] K. S. Thorne. Multipole expansions of gravitational radiation. *Reviews of Modern Physics*, 52:299–339, 1980. DOI: 10.1103/revmodphys.52.299 23

[136] K. S. Thorne. Gravitational waves. In *Particle and Nuclear Astrophysics and Cosmology in the Next Millennium, Proceedings*, pp. 0160–184, Summer Study, Snowmass, June 29–July 14, 1995. DOI: 10.1088/978-0-7503-1393-3ch1 62

[137] S. E. Timpano, L. J. Rubbo, and N. J. Cornish. Characterizing the galactic gravitational wave background with LISA. *Physical Review*, D73:122001, 2006. DOI: 10.1103/phys-revd.73.122001 88

[138] M. Tinto and S. V. Dhurandhar. Time-delay interferometry. *Living Reviews in Relativity*, 8(4), 2005. DOI: 10.12942/lrr-2005-4 68, 71

[139] R. van Haasteren et al. Placing limits on the stochastic gravitational-wave background using European pulsar timing array data. *Monthly Notices of the Royal Astronomical Society*, 414(4):3117–3128, 2011. DOI: 10.1111/j.1365-2966.2011.18613.x [Erratum: *Monthly Notices of the Royal Astronomical Society*, 425(2):1597, (2012)]. 74

[140] R. van Haasteren and Y. Levin. Gravitational-wave memory and pulsar timing arrays. *Monthly Notices of the Royal Astronomical Society*, 401:2372, 2010. DOI: 10.1007/978-3-642-39599-4_3 91

[141] R. van Haasteren, Y. Levin, P. McDonald, and T. Lu. On measuring the gravitational-wave background using pulsar timing arrays. *Monthly Notices of the Royal Astronomical Society*, 395:1005, 2009. DOI: 10.1111/j.1365-2966.2009.14590.x 74

[142] I. Vega, B. Wardell, and P. Diener. Effective source approach to self-force calculations. *Classical and Quantum Gravity*, 28:134010, 2011. DOI: 10.1088/0264-9381/28/13/134010 28

[143] J. P. W. Verbiest et al. The international pulsar timing array: First data release. *Monthly Notices of the Royal Astronomical Society*, 458(2):1267–1288, 2016. DOI: 10.1093/mnras/stw347 91

[144] A. Villenkin and E. P. S. Shellard. *Cosmic Strings and Other Topological Defects*. Cambridge University Press, 2000. 90

[145] M. Volonteri, T. Bogdanovic, M. Dotti, and M. Colpi. Massive black holes in merging galaxies. arXiv:1509.09027, 2015. DOI: 10.1017/s1743921316005366 84

[146] V. P. Frolov and I. D. Novikov. *Black Hole Physics, (Basic Concepts and New Developments)*. Springer Netherlands, 1998. DOI: 10.1063/1.1292486 3, 4

[147] S. J. Waldman. The advanced LIGO gravitational wave detector. In *22nd Rencontres de Blois on Particle Physics and Cosmology*, Blois, Loire Valley, France, July 15–20, 2011. 61, 66

[148] Y. Wang and S. D. Mohanty. Particle swarm optimization and gravitational wave data analysis: Performance on a binary inspiral testbed. *Physical Review*, D81:063002, 2010. DOI: 10.1103/physrevd.81.063002 51

[149] B. Wardell and A. Gopakumar. Self-force: Computational strategies. *Fundamental Theories of Physics*, 179:487–522, 2015. DOI: 10.1007/978-3-319-18335-0_14 29

[150] C. M. Will. The confrontation between general relativity and experiment. *Living Reviews in Relativity*, 9(1):3, March 2006. DOI: 10.12942/lrr-2006-3 1

[151] C. M. Will and A. G. Wiseman. Gravitational radiation from compact binary systems: Gravitational wave forms and energy loss to second post-Newtonian order. *Physical Review*, D54:4813–4848, 1996. DOI: 10.1103/physrevd.54.4813 23

Author's Biography

STANISLAV BABAK

Stanislav (Stas) Babak graduated from Moscow State University. He has a Ph.D. in Relativistic Astrophysics from Cardiff University. He worked for 13 years in Germany at the Max-Planck Institute for Gravitational Physics (Albert Einstein Institute). After Germany, he moved to France taking the position of *Directeur de Recherche* at CNRS in the laboratory AstroParticle and Cosmology. He is a member of LISA consortium, European Pulsar Timing collaboration. He has joined LIGO collaboration in 2001 and moved to Virgo after moving to France.

Printed in the United States
by Baker & Taylor Publisher Services